Powering the Green Shift: Optimizing Lithium-Ion Batteries for Sustainable Tools

Faris

Contents

Contents

1 Introduction

With the rise in popularity to use more sustainable products [1] [2], companies have been turning to use Lithium-ion batteries in place of gasoline to power their products. In the development of using this relatively new battery technology, companies must adapt their designs to maximize the potential power of these batteries.

One such company is Globe Group AB, who's vision is to lead the world away from fossil fuels and rely upon renewable energy. They are working towards this goal by developing all their tools with rechargeable battery packs ranging from hand drills and chainsaw to lawnmowers and leaf blowers. These battery packs have a range of voltage and capacity in accordance with the product they are used for. This book focuses on the thermal behavior of the 82V220 6Ah rechargeable battery pack.

For the duration of the project David Samvin, a Professor from Jonkoping University, is supporting and supervising the approach and research end of the book. To represent Greenworks Kim Larson and Mike Heath are assigned support the engineering and design side.

1.1 Background

Globe Group AB is the parent company of Greenworks, Power-works, and Cramer. Greenworks is the base consumer level of outdoor products, Power-works is the high-end consumer brand, and Cramer is the professional brand. All three brands share product engineering and design of their products but have different consumers and regions around the world they sell to. By having different brands, Globe Group can sell similar products with different price points and levels of power without having to re-engineer an entire product.

The benefits of Globe Group's all-electric tools line, besides sustainability, is that all these products are more efficient, produce zero emissions, are much quieter, easier to maintain, contain fewer parts to replace, are much easier to start, and perform on par with their gasoline counterparts. The main drawback of these battery-powered tools is the runtime and the recharge time of the battery packs.

The battery pack Cramer produces is the 82V220 6Ah battery pack, a 3.1 kg removable battery pack for the 82-volt power tool line. The battery pack itself contains 40 lithium-ion 18650 batteries that together can produce 2.1 kW of power, listed from the Greenworks specifications [3]. While the battery pack can meet the power requirements for most of the tools in the product line, there are few cases where the battery pack falls short. During extended and high-power operations, such as cutting thick, dense logs with a chainsaw, the battery pack draws higher currents, over 30 Amps. The high current draw causes the battery pack to rapidly heat up and eventually hit the set safety cutoff temperature of 75 °C, at which the battery pack cannot be charged or used in the tool until it cools down to a safe level of 40 °C [3] [4].

The overheating of the battery pack may cause many interruptions during work, and significantly increases the recharge time. From a consumer point of view, having work interruption or having to purchase double the amount of battery packs is frustrating and hurts the reputation of the battery-powered tool line. To address this issue, Globe Group has set the following objectives for the next generation of battery packs:

- Continually discharge at 35 Amps without internal temperature reaching 75°C.
- Non-continually discharge at 55 Amps without internal temperature reaching 75°C.
- Be able to provide 3.5 kW of power safely.
- Maintain an ingress level of IPX4.
- The non-common outer surfaces of the battery must not reach 55°C.
  - o The outer surfaces of the plastic casing of the battery pack consider as non-common surfaces.
- The common outer surfaces of the battery must not reach 45°C.
  - o The magnesium top cover of the battery pack considers as common surfaces.
- Reduce the weight of the battery pack.
- Reduce the cost of the battery pack.
- Interface with the current line of tools.

1.2 Purpose and research questions

During the product's development process, the aesthetics and design of the product is always changing to improve upon the previous product, include more features, and decrease the manufacturing costs. Throughout this product development, it is a task of its own to manage all the features and engineering requirements of the product.

To help Greenworks in their product development, the next generation of battery packs, this book focuses on creating a thermal simulation and investigate cooling suggestions for the battery pack. In the list of objectives for the battery pack, a significant issue is managing the heat it produces. While this overheating issue can be solved in many ways, the fastest, most affordable way is to compare different cooling solutions in a simulation. Greenworks up to this point has not utilized simulations for thermal testing their products, so in addition to aiding in the development of the next battery pack, this would be a new tool for their research and development team to use for future projects.

The results of this investigation answer:

- Why is the battery pack overheating?
- What is the procedure to conduct credible thermal testing on this battery pack?

- What are the simulation parameters to recreate the results collected from the battery discharge temperature testing?
- Can the parameters of the simulation be used for other tests or projects?
- What are some suggestions for the redesign to meet the thermal requirements?

The approach to developing the simulation parameters for the battery pack is to understand the materials and construction of the battery pack, model the pack in CAD, and load the model with the appropriate amount of thermal energy. The results from this simulation are compared and calibrated to real-life tests until the simulation results match what is happening in real-life. Once the simulation results are satisfactory, then new designs and cooling solutions can be tested and compared to the current design.

1.3 Delimitations

Greenworks has set many objectives for the next generation of the battery pack that is quite large and broad, considering the time and resources of this book; the scope is limited to not considering the following:

- Creating prototypes of new designs.
- Creating detailed CAD models of new design ideas.
- Creating the redesigns for manufacturability.
- Estimating the manufacturing cost associated with each new design concept.
- How to market the new design.
- Packaging and shipping requirements.
- The tests do not consider the product in extreme external environmental.
- This book study does not investigate the battery pack design for impact, ingress protections, or stress tests on the current or new concepts of the battery pack.
- The study does not consider using alternative power sources or changing the number of battery cells inside the battery pack.

In addition, due to the global pandemic of the COVID-19 virus, the project budget from Greenworks was reduced, and quarantine restrictions limited the battery testing. As a result of the budget, fully certified testing standards could not be purchased, and the tests were limited to use only 9 external sensors. The quarantine restrictions reduced the number of tests that could be performed, limiting the testing to only 35- and 45-Amps tests.

1.4 Outline

This book reviews the theories that are required to understand the Lithium-ion battery, heat transfer, and how simulations are for product development. Then move on to how the battery experiments were conducted, how data is collected, and how thermal simulations are set up. Finally, the parameters selected for the final simulation are explained, and the results from both the discharge tests and thermal simulations are analyzed.

2 Theoretical background

This section of the report introduces the theoretical backgrounds that help understand the rest of the book. These theories were used both during the experimental and simulation to understand and solve the heat transfer problems. This chapter explains the theory behind the lithium-ion batteries and the thermal behavior of the batteries during charging and discharging, to identifying the heat that generates under the working process of the batteries and numerical solution to solve this kind of problems. For a deeper understanding of the testing results and discussion, refer to the section about heat transfer.

2.1 Lithium-ion Batteries

Lithium-ion batteries are known as a high-density source of power because of high capacity, high efficiency, and long life. That is the reason Lithium-ion batteries became the most desirable source of power for a wide range of devices and electric vehicles (EVs). With the increasing use of advanced battery-powered products, the demand for lighter and more compact battery packs is increasing, and this is where Lithium-ion batteries fit in, becoming more critical for product development [5].

In addition to these requirements, battery performance has also received much attention because it increases the product's range and performance, which is an essential goal in the industry. In response to all these needs, in 1991, Sony released its first commercial lithium-ion rechargeable battery.

Lithium-ion batteries divide into two different categories, such as primary (single-use) and secondary (rechargeable).

The lithium-ion batteries structure is comprised of a few main components, as shown in *Figure 1*. Electrodes (anode and cathode), current collectors, which are connected to each electrode. The separator separates two electrodes from each other and acts as a guard and only allows lithium ions to pass through. The separator is an insulator layer to avoid a short circuit between anode and cathode. The separator layer is permeable for the lithium-ions because of its microporosity [6] [7].

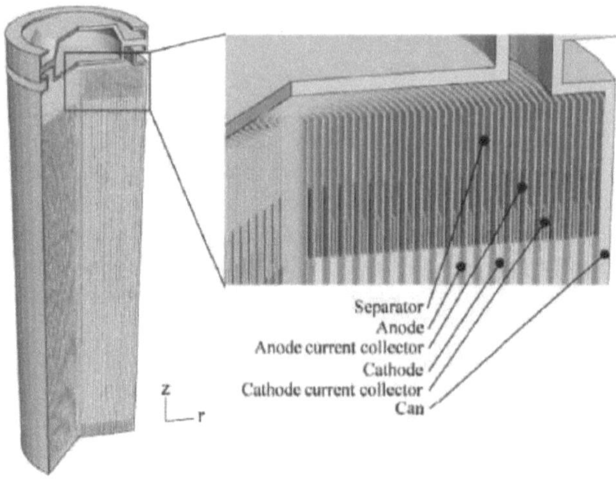

Figure 1: Basic structure of the lithium-ion cell [6]

During the charging and discharging of the batteries, lithium ions Li+ move between the anodes and cathodes due to the electrochemical reactions, shown in *Figure 2* [8].

$$xLiC_6 + Li_{1-z}Mn_2O_4 + Li_{1-y}Ni_{1/3}Mn_{1/3}Co_{1/3}O_2$$

$$\underset{charge}{\overset{discharge}{\rightleftharpoons}}$$

$$Li_zMn_2O_4 + Li_yNi_{1/3}Mn_{1/3}Co_{1/3}O_2 + C_6$$

Figure 2: Overall cell reaction, where x=y+z, 0<y≤1 and 0<z≤1

These reactions can occur at both electrodes, depending on the charging and discharging of the batteries, that is the reason that makes it possible for the secondary batteries to recharge, as shown in *Figure 3* [9] [7].

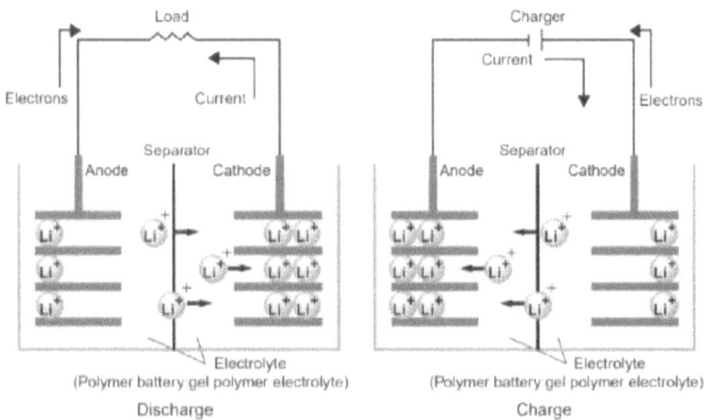

Figure 3: charge and discharge mechanism of lithium-ion rechargeable batteries.

2.2 Charging and discharging of the lithium-ion batteries

The condition of charging and discharging of the batteries has a direct impact on the battery's life and life cycle, thus in the performance of the batteries.

2.2.1 Depth of discharge

The depth of discharge (DoD) can be crucial for the lithium-ion battery life cycle, therefore a partial discharge reduces the voltage and thus extends the battery life, and on the other hand, the high current causes greater DoD and elevated temperatures that reduce the battery life [8].

Table 1 shows how DoD affects the battery cycles of the standard cobalt-based lithium-ion batteries [10].

Depth of Discharge	Discharge cycles	
	NMC	LiPO4
100% DoD	~300	~600
80% DoD	~400	~900
60% DoD	~600	~1500
40% DoD	~1000	~3000~
20% DoD	~2000	~9000
10% DoD	~6000	~15000

Table 1: Cycle life as a function of depth of discharge

2.2.2 **Charging**

The charging level of batteries is as important as the DoD in battery life and cycle count. In general, most lithium-ion batteries can be charged to 4.2V / cell. Most companies want to use the highest voltage because it provides higher capacity, but the high charge per cell reduces the battery life. Therefore, battery experts recommend using 3.92 V / cell as the optimal charge voltage to omit all voltage-related stresses [10] [11].

A study by Chalmers University shows that each 0.10 V drop below 4.2V / cell for doubles the cycles and higher voltage than 4.2V / cell can reduce battery life, as it shows in *Table 2*, [8] [12].

Charge level (V/cell)	Discharge cycles	Available stored energy
4.30	150-250	110-115%
4.25	200-350	105-110%
4.20	300-500	100%
4.15	400-700	90-95%
4.10	600-1000	85-90%
4.05	850-1500	80-85%
4.00	1200-2000	70-75%
3.90	2400-4000	60-65%

Table 2: Effect of charge voltage limit to discharge cycles and capacity.

The effect of charge voltage on cycle performance shows in *Figure* 4.

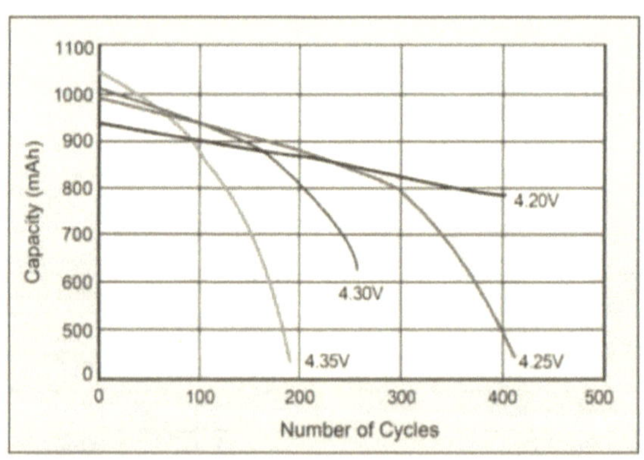

Figure 4: the voltage charge effects on the cycle life

2.2.3 **Battery State of Charge**

Batteries of all kinds are like fuel tanks where they have a limited amount of potential energy, and they can run till empty, but unlike a fuel tank in a car, it is tricky to measure how much energy is in a battery. The first thought is to compare the battery's voltage to its application. This method is suitable but can be misleading since the battery voltage is not loaded when being measured and does not show what percentage of the battery is remaining, also known as the batteries state of charge (SOC). One of the standard methods to measure the SOC is the *Coulomb Counting Method,* where instead of measuring the energy left in the battery, you measure the amount of energy that has left the battery [13]. This method is right for constant current discharge testing, whereas the method is named counting the coulombs, as shown in *equation (1)*

$$\text{SOC} = \text{SOC}(t_0) + \frac{1}{C_{rated}} \int_{t_0}^{t_0 + \tau} (I_b - I_{loss}) \, dt \tag{1}$$

Where $\text{SOC}(t_0)$ is the initial SOC, the C_{rated} is the maximum capacity of the battery being tested, I_b is the battery current, and τ is the total time, I_{loss} is the current consumed by the loss reactions.

2.3 **Thermal Behavior of Lithium-ion Batteries**

Lithium-ion batteries contain high amounts of energy and generate a significant amount of heat during charging and discharging. The heat generation of lithium-ion batteries divides into two main parts, reversible heat generation q_{re} and irreversible heat generation q_{irr}. This heat is directly related to the discharge rate of the batteries; the irreversible heat increases with a high discharge rate, while reversible heat is more dominant at lower discharge rate [14]. The total heat generated by the battery can be calculated by following *equation (2)* [15].

$$q_{tot} = q_{re} + q_{irr} \tag{2}$$

The reversible heat is related to change in entropy of electrodes during charging and discharging due to the intercalation and de-intercalation processes. Changes in entropy in the electrodes depend on the electrode's materials and the condition of charge. The state of charge (SOC) is known as a factor that has a great impact on reversible heat [16]. The reversible heat calculation is shown in the following *equations (3)&(4):*

$$q_{re} = S_{(a,i)} \, j_{(loc,i)} \, \frac{T \Delta S}{n \cdot F} \tag{3}$$

$$\Delta S = n \cdot F \; \frac{\delta U_i}{\delta T} \tag{4}$$

Where $S_{a,i}$ is the specific surface area (m^{-1}), $j_{loc,i}$ local current density (Am^{-2}), ΔS is the change in entropy, U_i is the open-circuit voltage which describes the equilibrium potential of electrodes as a function of SOC, F is Faraday's constant $(=96487 \; C \; mol^{-1})$, n is number of moles of electrons.

Irreversible heat generation is divided into two different categories, Ohmic heat generation that occurs due to internal ohmic resistance of the battery and active polarization heat generation due to electrochemical reaction that occurs during the charging and discharging process [14] [17]. The irreversible heat is calculated by following *equation (5)*.

$$q_{irr} = q_p + q_{ohm} \tag{5}$$

The polarization heat generation is calculated with the following *equation (6)*:

$$q_p = S_{(a,i)} \cdot j_{(loc,i)} \left(\varphi_{(1,i)} - \varphi_{(2,i)} - U_i \right) \tag{6}$$

U_i is open-circuit voltage, which describes the equilibrium potential of electrodes as a function of SOC. Since temperature effects on the batteries discharge, process a Taylor expansion performed at the temperature reference, *equation (7)* [15].

$$U_i = U_{ref,i} + (T - T_{ref})\frac{dU_i}{dT} \tag{7}$$

The ohmic heat is due to the internal resistance; this method of heating is quite common in consumer products such as toaster, electric stovetop, and microprocessors in computers. The heat generated by these devices are solely based upon the current and the resistance of the device. Ohmic heating in a battery is based on three different factors, material resistance, the movement of ions through the electrolyte, and the heat generate through the current collector [14]. The ohmic heat can be calculated through the following *equation (8)*:

$$q_{ohm} = I^2 R \tag{8}$$

Where I is the discharge current, and R is the internal resistance of the cell. The interest in running batteries at high current is for the high-power output. Because of this interest in the higher power output, the thermal behavior of batteries becomes more critical, since the higher current discharge dramatically increases the rate of overheating. Research and experiments have found that the battery's performance and lifecycle are highly dependent on the temperature condition. So, temperature variation must be handled inside the battery to avoid exceeding the heat, thus avoiding thermal runaway in the battery, which in the worst case can also cause fire or even explosion. Therefore, it is crucial to be able to handle the temperature to keep the product at its optimal operating temperature.

Therefore, it is recommended not to use lithium-ion batteries outside of the specified temperatures, as batteries do not work correctly under these conditions. Furthermore, these temperatures can lead to permanent or high damage to the cells. That is why thermal management is so crucial to the battery's performance and lifetime [18] [7].

2.4 Heat Transfer

Heat transfer is the movement of thermal energy between different mediums. Heat itself is the vibration molecules within the medium; the movement of these molecules is measured in terms of temperature using the units such as Celsius, Fahrenheit, and Kelvin. For heat transfer to occur, there must be a temperature difference between the objects in question. Heat is always moving from the object with a higher energy level to the object with a lower energy level. There are three different mechanisms for heat transfer, conduction, convection, and radiation [19] [20].

2.4.1 Conduction

Conduction is the process where heat is transmitted through a solid object or between two solid objects in direct contact with each other. The rate at which the heat is transfer is based upon the difference in temperature and the material's thermal conductivity. All types of matter have value for thermal conductivity, and this value represents how well the matter can transfer energy by the diffusion process [19]. Heat transfer through conduction can be expressed with Fourier's Law seen in *equation (9)*.

$$Q_{Cond} = -kA\,\frac{dT}{dx} = kA\,\frac{(T_1 - T_2)}{L} \qquad (9)$$

Where k is the materials thermal conductivity (W/m K), dx is the material thickness, and A is the surface area the heat is being transmitted through. Fourier's equation can also be re-arranged into *equation (10)*.

$$Q_{Cond} = \frac{(T_1 - T_2)}{R_{wall}} \qquad (10)$$

$$R_{wall} = \frac{L}{kA} \qquad (11)$$

R_{wall} in *equation (11),* is the thermal resistance through a material. In this form, the equation can be used to calculate the heat transfer through several layers of material or used simply calculate the amount of thermal resistance through a set of materials.

2.4.2 **Convection**

Convection is the transfer of heat through the movement of fluid, which can be either liquid or gas. The surface in contact with the fluid will heat the nearby fluid molecules, which will then move away from the surface due to the flow of the fluid or gravity. The value for thermal coefficients is based on the fluid properties, surface geometry, and temperature of the surface. When you have a hot surface, the rate at which it loses heat to the air is much high than when the same surface is only a few degrees colder or warmer to the air; this is because the hotter surface can heat the air molecules faster which causes their density to decrease and begin to rise making space for cooler molecules to repeat the same process. This heat transfer is described in *equation (12)*, where the difference in surface, ambient fluid temperature is multiplied by the thermal coefficient (h).

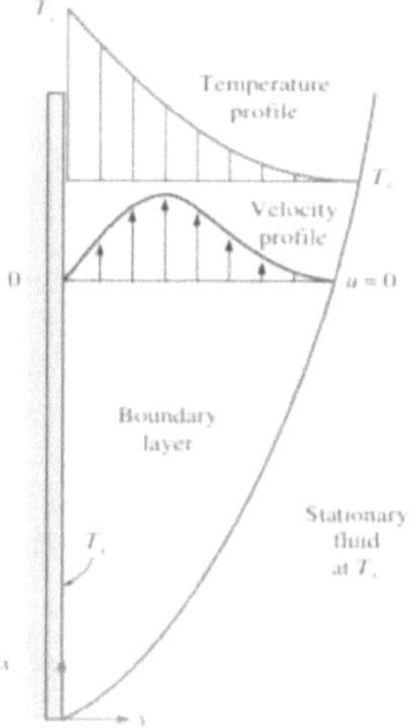

Figure 5: diagram of convection on vertical surface.

$$q_n = h(T_s - T_\infty) \tag{12}$$

The thermal coefficient *(h)* can be complicated to calculate for a given situation and changes based on the temperatures in the system. For quick reference, *Table 3*, recreated from [21] shows the range of thermal coefficient that can be applied for a given situation.

Typical Values of the Convection Heat Transfer Coefficient	
Process	$H\left(\dfrac{W}{m^2 K}\right)$
Free Convection	
Gases	2-25
Liquids	50-1000
Forced Convection	
Gases	25-250
Liquids	100-20,000
Convection with phase change	
Boiling or Condensation	2500-100,000

Table 3: Typical Values of the Convection Heat Transfer Coefficient.

While *Table 3* provides a sense of how drastically the coefficient can change between moving and free-flowing fluids, the range of values is so broad that a single value cannot be accurately used for a given situation. For a closer figure *equation (14)* is used, this formula uses several essential relationships between fluid properties, surface shapes, and flow types.

$$Nu = \frac{hL_c}{k} \tag{13}$$

$$h = Nu\frac{k}{L_c} \tag{14}$$

Where h is the thermal coefficient, L is the characteristic length, k is the thermal conductivity of the fluid. *Nu* is the *Nusselt number,* a dimensionless number that varies based on the type of convection occurring and shape of the surface. The *Nusselt Number* is derived from the ratio of the heat transfer equations of convection and conduction shown in *equation (17)*.

$$\dot{q}_{conv} = h\Delta T \tag{15}$$

$$\dot{q}_{cond} = k\frac{\Delta T}{L} \tag{16}$$

$$\frac{\dot{q}_{conv}}{\dot{q}_{cond}} = \frac{h\Delta T}{k\Delta T/L} = \frac{hL}{k} = Nu \tag{17}$$

From *equation (17)*, it is evident that as the *Nusselt number* rises, the more effective the convection becomes. [21] when the *Nusselt number* is equal to 1, the fluid layer behaves the same as pure conduction.

The method to calculate the *Nusselt number* in natural convection over a surface is based on the *Rayleigh Number Ra* and the geometry of the surface. Depending upon the surface position and shape, these equations vary, in the appendix section 7,2

Table 12 shows the calculation method for horizontal and vertical plates in natural convection, which both depend on *Rayleigh number*. The *Rayleigh number* is the product of the *Grashof equation (19)* and *Prandtl numbers equation(18)* shown in *equations (20)*. The *Grashof number* is another dimensionless number that represents the ratio of the buoyancy force and viscous force acting on the fluid.

$$Pr = \frac{\mu C_p}{k} \tag{18}$$

$$Gr_L = \frac{g\beta(T_s - T_\infty)L_c^3}{v^2} \tag{19}$$

$$Ra_L = Gr_L * Pr = \frac{g\beta(T_s - T_\infty)L_c^3}{v^2} * Pr \tag{20}$$

Where:

g = gravitational acceleration, $m/_{s^2}$

β = coefficient of volume expansion, $1/K$ ($\beta = 1/T$ for ideal gases)

v = The kinematic viscosity of the fluid, $m^2/_s$

The Prandtl number and the kinematic viscosity of air at 1 atm are listed in the fluid properties *Table 14* found in appendix section 7,4 [22].

The calculation for convection heat transfer can go quite deep into the fluid properties and surface geometry, which is why when determining which *Nusselt* equations to use, it is crucial to identify the type of system that will be analyzed before any calculations.

2.4.3 **Radiation**

Compared to conduction and convection radiation, heat transfer one of the harder types of heat transfer to visualize. Thermal radiation is energy continuously emitted through electromagnetic waves; this energy is emitted by everything that is above zero degrees, Kelvin. This form of heat transfer can travel through any medium but has the best transfer in pure vacuums. Every object emitted energy in every direction the form of photons that hit and move other molecules. An example of radiation heat transfer is the heat we obtain from the sun. Heat transfer through this method can be calculated by two different *equations (21)*and *(22).*

$$q_{rad} = h_r \, A \, (T_s - T_\infty) \tag{21}$$

$$q_{rad} = \varepsilon \, \sigma \, (T_s^4 - T_\infty^4) \tag{22}$$

Equation (21) is based on the surface area and the coefficient of radiation h_r which is based on the surface emissivity of the material and *Stefan-Boltzmann* constant seen in *equation (23).* Where σ is *Stefan-Boltzmann* constant ($5.670400 \times 10^{-8} \, \frac{w}{m^2 K^4}$).

$$h_r = \varepsilon\sigma(T_s + T_{sur})(T_s^2 + T_{sur}^2) \tag{23}$$

Equation (22) derived from *Stefan-Boltzmann* Law simply uses the surface emissivity *Stefan-Boltzmann* constant to calculate the heat transfer. The surface emissivity is based on the concept of the *Blackbody principle,* which states that a perfect *Blackbody* can emit and absorb radiation entirely. As the name states, objects with darker surfaces can absorb and emit much more radiation than light-colored surfaces. The surface emissivity is on a 0 to 1 scale where a blackbody object can have an emissivity of 1 and surfaces that are very reflective such as polished nickel and aluminum have values as low as 0.03. *Table 13* in the appendix section 7.3 displays the emissivity values of a few different materials and shows that as the temperature increase, the emissivity value of the materials increases as well.

2.5 Heat Capacity and Specific Heat

An additional method for measuring the energy loss due to heat in a system is through calculating the heat capacity. This method is based on three elements the mass of the material, the material's specific heat value, and the change in temperature, which creates the simple *equation (24).*

$$q = mc\Delta T \tag{24}$$

Where q is the energy required to raise or lower the temperature (ΔT) of a material with a given mass (m). The specific heat of a material represents the amount of energy required to raise the temperature one kilogram of the given material by one-degree Kelvin. This method for calculating energy loss due to heat is best using in a device called a calorimeter. This device is an extremely insulated container filled with water that measures the change of water temperature as the object being test heats up.

2.6 Electrical Circuits

When dealing with any kind of electrical system, *Ohm's Law* must be considered and can be found in most physics and electrical textbooks [23]. *Ohm's Law* describes the relationship between the voltage, current, and resistance in an electrical system. The voltage drops across a given electrical component is equal to the product of the current and resistance of the given electrical component, written out as *equation (25).*

$$V = IR \tag{25}$$

$$I = \frac{V}{R} \tag{26}$$

$$R = \frac{V}{I} \qquad (27)$$

Equations (26) and (27) are a rearrangement of equation (25) that solve for current and resistance in a given system. To calculate the individual values for voltage, current, and resistance, one must consider the circuit type. Series and parallel are the basic circuit types displayed in *Figure 6.*

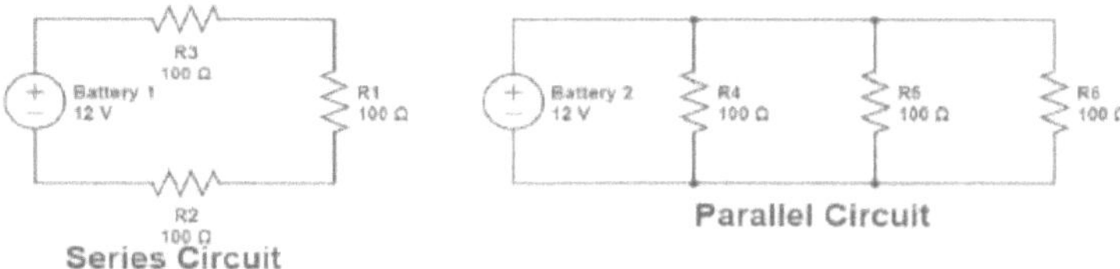

Figure 6: Series and parallel circuit configurations.

Series circuits have the same current at every point in the circuit, but the voltage drops at each resistor in the series. This voltage drop is calculated with *ohm's law*, where the total voltage is the sum of all the drops in the system.

For parallel circuits, the voltage will remain the same between the parallel set of resistors, but the current splits between each resistor. The current varies depending on the resistance of the resistor in each parallel set.

The relationship for calculating the resistance in different areas of series and parallel circuits depends on where it's measured. In a series circuit, the total resistance is calculated using *equation (28)*. For parallel circuits, the total resistance in the system is calculated with *equation (29)*.

$$R_T = \sum R_n = R_1 + R_2 + \dots + R_n \qquad (28)$$

$$\frac{1}{R_T} = \frac{1}{R_1} + \frac{1}{R_2} + \dots + \frac{1}{R_n} \qquad (29)$$

2.7 Testing Standards

Testing standards are guides for others to use to conduct the same test in their environment. Tests are used to determine the safety of a product or to compare different materials or products to each other. The standards are typically created and controlled by government agencies or testing companies such as Underwriters Laboratories (UL) or International Organization for Standardization (ISO). Often these organizations have similar tests with slightly different procedures to test a product. Depending upon where a product is being sold, it must meet minimum safety standards for that region to open to the public for purchase.

The testing standards have detailed procedures for conducting tests such that the results are recognized by government agencies or other researchers as valid results. When using one of these standards, it is imperative to clearly understand the purpose of the test and how the results are collected so that all outside influences cannot compromise the validity of the results.

These testing agencies are continually creating new tests and refining old ones as technology advances and new safety issues arise. There may not be a formal standardized test available for a given situation; in this case, new tests must be presented and agreed upon, like *Daniel Doughty* [24] proposal for battery failure propagation testing. Which showed none of the current thermal runaway battery packs tests, test for propagation failure throughout the battery pack, which has been recorded to have happened in Dell Laptops. When there are no standardized tests for the given situation, companies create their testing standards for testing their own or competitors' products.

2.8 Simulations

A significant part of engineering is being able to predict how designs will perform before manufacturing or testing anything. One of the many tools that assist engineers with their predictions is to use finite element analysis with help of computer aided engineering simulation software (CAE). Simulation software are commonly paired with CAD modeling software, allowing the user to move the CAD model into the simulation environment seamlessly. Here the model can be tested under different conditions it may undergo in its life cycle such as bending stress, wear, or heat transfer. Accuracy of the simulations between real life and the computer all depends on how well the simulation parameters are entered. The simulation parameters include the material properties, loads on the model, and boundary conditions. The values for these parameters, such as material properties, can be found in online databases or manufacturing companies, other values may need to be calculated or measured based on the condition of the real-life environment such as temperature or heat power. When used correctly, simulations results can show how the design acts under different loading conditions. The value of using simulations is that new concept can be easily tested, much less effort is required to conduct tests, and the simulations can provide detailed, accurate results all of which saves time and money.

2.8.1 **Solidworks simulation**

The simulations preformed in this book are using conduction and convection in a transient study. Solidworks uses following governing equation to solve the transient heat transfer in the simulations [25].

$$\frac{\partial}{\partial x}\left(k_x \frac{\partial T}{\partial x}\right) + \frac{\partial}{\partial y}\left(k_y \frac{\partial T}{\partial y}\right) + \frac{\partial}{\partial z}\left(k_z \frac{\partial T}{\partial z}\right) + \dot{q} = pc\frac{\partial T}{\partial t} \tag{30}$$

This formula expresses the rate of the heat in the three direction of the cartesian coordinate system. Where k is the coefficient of heat transfer, p is the density of the material and c is the specific heat of the material. For 3-D models the direction of the heat transfer is important since the material properties and geometry influences the rate of heat in their respective direction. Since most material thermal properties are orthotropic the thermal conductivities are equivalent in every direction. This is one of the reasons why simulating a single part with one material is faster to calculate. In the case for the battery pack with complex geometry and many different materials, the simulation must adjust the heat transfer coefficient for each material [25].

$$T(x,y,z,t) = T_0 \ for \ \ t > 0 \ on \ S_1 \tag{31}$$

$$k_x \cdot \frac{\partial T}{\partial x} \cdot l_x + k_y \cdot \frac{\partial T}{\partial y} \cdot l_y + k_z \cdot \frac{\partial T}{\partial z} \cdot l_z + q = 0 \ for \ t > 0 \ on \ S_2 \tag{32}$$

$$k_x \cdot \frac{\partial T}{\partial x} \cdot l_x + k_y \cdot \frac{\partial T}{\partial y} \cdot l_y + k_z \cdot \frac{\partial T}{\partial z} \cdot l_z + h(T - T_\infty) = 0 \ for \ t > 0 \ on \ S_3 \tag{33}$$

To solved equation 30, two boundary conditions must be specified. The first boundary condition is based on the first law of thermodynamics where no energy can be created or destroyed. In the simulation all the heat between being generated and dispersed must be accounted for. The second boundary condition is based upon when the heat transfer changes from within the system to the surrounding environment. If the surrounding temperature is lower than the system, the heat will be dissipated to the surrounding environment via convection.

Lastly initial condition must be set. In this case the initial are the starting temperatures of each material in the system including the surrounding air.

$$T(x,y,z,t = 0) \ \bar{T}_0(x,y,z) \ in \ V \tag{34}$$

2.8.2 **Knowledge of simulation**

Understanding the how Solidworks solves this heat transfer problem helps to set up the simulation for the battery pack. One example is setting up the simulation for the battery pack with the casing on the battery.

When creating the model of the battery pack with the casing on, a solid part representing air gap between the battery cells and the outer wall must be created. The simulation can only transfer heat through conduction which requires two solid parts. When heat is transferred via convection the heat is reduced to the surrounding temperature and cannot be transferred to another object. Therefore, the heat between the battery cells and the outer wall cannot be transferred via convection as it does in real-life. To overcome this issue the air must be represented as a solid material, transferring the heat to the outside of the case via conduction. Normally heat transfer through air would be convection but since the battery was sealed with an IPX4 rating it is safes to assume the all the heat in the air transfer directly to the outer casing. The IPX4 rating means there is little to no air flow to the outside of the battery pack, so the air can be treated as a solid material transferring the heat via conduction.

Ideally, there would be no need to perform any real-life testing, and the new design concepts could be simulated straight away. The issue is that the results of the simulation may not match with what is happening in real-life. Simulating heat transfer is relatively straight forward to set-up and run, but many outside factors can alter the simulated result from the real-life result, for instants:

- Most materials are not homogeneous in their material properties.
- Every component in contact with the battery pack can influence the results, including the temperature sensors.
- The geometry of the virtual model will not precisely match the real product.
- It is challenging to estimate the environmental factors that can influence the results.
- Some of the real-life values are difficult to calculate and measure (heat power)

With all these factors to consider, it is unreasonable to create a simulation that can exactly match the real-world conditions without comparing the simulation to a real-life test.

The simulation must be compared to the real-world test to validate and gauge the accuracy of these simulation results. To effectively utilize the power of the simulation software, there must be a baseline for comparing the simulation to a real-life test. The comparison is made by testing multiple battery packs under the same controlled conditions and while recording temperature results from the same locations. This temperature data is the target for the simulation to match.

2.9 Optimization

Optimization is a process to achieve the optimal solution for a problem under the given circumstances. By using the mathematical principles and methods to maximize desired factors such as efficiency, productivity, strength, and minimizing the undesired factors such as cost and risk. Design engineers are using optimization to find the optimal design by using the design constraints and criteria.

3 Method

3.1 Product Analysis

The battery pack under investigation is the Cramer 82V6AH. The shape and contact interface of the Cramer battery is shared between the 60, 80, & 82V battery packs and the other two brands Greenworks and Powerworks. The difference between the battery packs is the number of cells and the configuration of the cells. For the Cramer 82V6AH, the specifications show that the battery uses 40 Lithium-ion 18650 cells in a 20S2P configuration. Meaning the 40 batteries have two parallel sets of 20 cells in series connection. A simplified version of this configuration is shown in *Figure 7*. This configuration matches the product performance specifications listed in *Table 4*. Between the three brands, this is the most powerful battery pack available with a capacity of 430 Wh. The Cramer 82V6AH is also has a weatherproof rating of IPX4, meaning it is protected against splashing water from all directions.

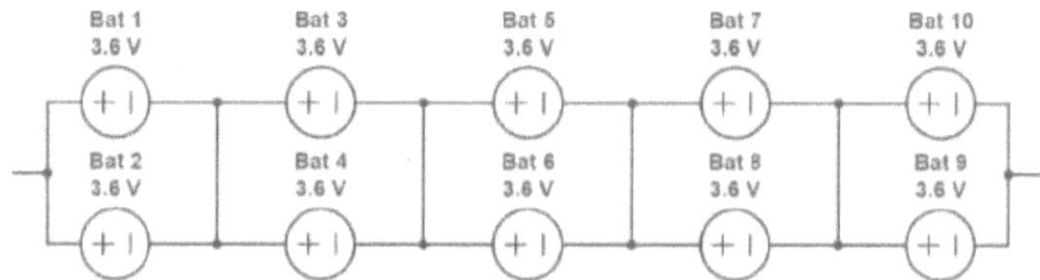

Figure 7: Series configuration of the cells – simplified.

Inside the battery, the first thing that is distinctly different from the rest of the battery packs in the product line are the aluminum heat sinks seen in *Figure 8*. Between each row of batteries, an aluminum heat sink is in contact with the battery cell for the sole purpose of absorbing the heat generated during discharge. A full teardown of the battery pack is not required since the arraignment of the cells was quite simple, and the full CAD assembly was available to review.

Figure 8: Positions of Aluminum heatsinks

Item	Specification	Remark
Charge limit voltage	84V max	4.2V/ Cell
Nominal Voltage	72V	3.6V/Cell
Discharge ending voltage	56V	2.8V/Cell – Power cut off by tools
Nominal capacity	6000 mAh	At 25°C
Battery Resistance	≤150 mΩ	AC 1 KHz after standard charge
Operating charge temp.	6~40°C	45%~85% Relative humidity
Operating discharge temp.	-14~45°C	
Cut off temperature	75°C ± 3°C	Cut off by tools

Table 4: Copied from Cramer 82V battery pack Specification sheet [3].

Item	Specification	Remark
Battery Type	Lithium-ion rechargeable battery cell: US18650VTC6	Sony Product
Rated Capacity	3000 mAh	
Nominal Capacity	3120 mAh	
Maximum Charing Voltage	4.25V	
Nominal Voltage	3.6V	
Discharge Cut-off voltage	2.5V	Recommend Voltage
	2.0V	Lower Limited Voltage
Continuous Maximum Discharge Current	30 A	With 80 C temp. cut off
	15 A	Without 80 C temp. cut off
Discharging above 30A limitations	30~40A	<40 Sec.
	~55A	<19 Sec.
	~80A	<6 Sec.
AC Impedance	8mΩ~18mΩ	After Standard Charging within 3 days
AC Impedance	8mΩ~18mΩ	Shipping Conditions
Allowable Environment Temperature	0~+60deg. C	Charging
	-20~+60deg. C	Discharging
Weight	46.6+-1.5 g	

Table 5: Specification of Lithium-ion cell - US18650VTC6 [9].

Figure 9: Interface of the battery pack.

3.2 Experiment

The basis for the testing is to analyze and map the rise of temperatures and the progression of heat transfer during discharge. The data from these tests are used to calibrate the simulations by comparing the simulation result versus the real-life testing results. The result from the simulation can also be used for later comparison with all the conceptual designs through design studies. The tests and simulations are used to help understand the transfer of heat and later can be used as a baseline for concept development of the new battery packs.

The testing method is based on preliminary research of the international testing standards and Greenworks testing standard for lithium-ion batteries. For lithium-ion batteries, many different standards are regulated by governments to protected individuals from the hazards associated with lithium-ion batteries. Therefore, many of these standards were studied to ensure the tests conducted with the Cramer battery pack would be safe and provide valid test results. Greenworks has testing procedures for many types of battery pack tests, including recording the battery temperature during discharge, which is precisely meant for this type of analysis. While these test procedures may be acceptable within Greenworks, there are no documents to support this experimental design. Research for the similar types of tests was conducted to crosscheck the Greenworks testing procedure, the list of tests that were researched are shown in *Table 6,* The research found international testing standards for thermal runaway, Electric Vehicle battery tests, and lithium cell abuse tests but no tests like Greenworks discharge temperature test. While none of the international tests were an exact match, they provided a useful reference for how a battery test should be conducted.

Organization	Test Number	Test Description
UL	9540A:2019	Standards for Thermal Runaway in Battery Energy Storage Systems
UL	2580	EV Li-Ion Battery Standards
UL	1973	Standard for Batteries for Use in Stationary, Vehicle Auxiliary Power and Light Electric Rail (LER) Applications
UL	840	Standard for Insulation Coordination Including Clearances and Creepage Distances for Electrical Equipment
UL	2271	Standard for Batteries for Use In Light Electric Vehicle (LEV) Applications
SAE	J2929	EV Li-Ion Battery Standards
SAE	J2464	Electric and Hybrid Electric Vehicle Rechargeable Energy Storage System (RESS) Safety and Abuse Testing
IEC	62660-3	EV Li-Ion Battery Standards
IEC	62984-2:20219	High-temperature secondary batteries - Part 2: Safety requirements and tests
IEC	60086-4	Safety of lithium batteries
IEC	60335-2-107	Particular requirements for robotic battery powered electrical lawnmowers
IEC	62798:2014	Industrial electroheating equipment - Test methods for infrared emitters
IEC	62446-3:2017	Outdoor infrared thermography Requirements for testing, documentation and maintenance for thermal abnormalities
IEC	62395-2:2013	Electrical resistance trace heating systems for industrial and commercial applications
IEC	62619	Safety requirements for secondary lithium cells and batteries, for use in industrial applications
IEC	62133	Secondary cells and batteries containing alkaline or other non-acid electrolytes
RTCA	DO-311	Minimum Operational Performance Standards for Rechargeable Lithium Batteries and Battery Systems
NASA JSC	20793 Rev D (2017)	Crewed Space Vehicle Battery Safety Requirements
SAND	2017-6925	Recommended Practices for Abuse Testing Rechargeable Energy StorageSystems (RESSs)
ECE	R100	EV Li-Ion Battery Standards
ISO	12405-3	EV Li-Ion Battery Standards
ASNT	1-57117-044-8	Nondestructive Testing Handbook, Third Edition: Volume 3, Infrared and Thermal Testing (IR)

Table 6: international testing standards.

3.2.1 **Testing Procedure for Lithium-ion Batteries**

The test procedure used for the Cramer battery pack is mainly based on the Greenworks temperature discharge test and supported with test procedures from the *NASA JSC 20793, SAND 2017-6925, UL 9540A:2019,* and *ECE R100* [26] [27] [28] [29]. While all these tests measure battery temperature, most involve either heating the battery from an external source or forcefully causing the battery to undergo thermal runaway.

The objective of the temperature discharge test is to monitor the change in temperature during discharge. The results are focused on the rate of temperature change and the location of the heat source. The tests were performed on 7 identical battery packs, provided by Greenworks, for a larger sample size and increase the validity of the results.

These tests were performed in the room seen in *Figure 10*, which was assumed to be a controlled environment based on the testing conducted in [30]. In this study, it was found that the effect of people moving inside of a large room with high ceilings has little to no effect on the airflow and convection value of the room. A controlled environment was essential to recreate the test environment in simulations and be able to compare the results to each other.

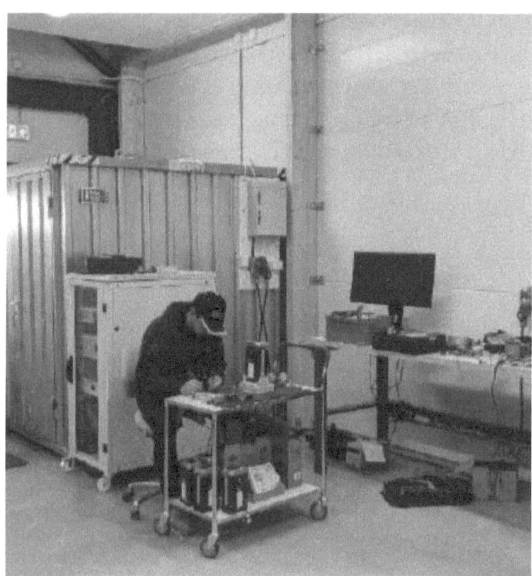

Figure 10: Testing environment.

The test has two main phases, first to identifying the source of heat, second to record the change in temperature of the heat source during the discharge of the battery pack. To conduct both parts required different measuring equipment but the same testing conditions.

Thermal Monitoring during discharge phase 1:

Description: Phase 1 test is to determine the location of the heat source within the battery pack. The location of the heat source is found with the use of a thermal imaging camera. The camera can provide a real-time image showing the location of where the heat is being generated. The battery pack must be at 100% SOC and allowed to acclimate to the testing temperature of 23° C ± 3°C for at least 3 hours after being removed from the recharge station. The pack is then connected to the discharge machine and discharged at the following rates:

- 35 Amps continuously
- 40 Amps continuously
- 45 Amps continuously
- Non-continuous current of 55 Amps (similar to real product use)

The tests are first conducted with the outer housing on to observe the battery pack as it is designed, then the tests are then repeated without the housing. This test may be conducted in parallel with phase two, once the location of the heat source is identified. The location of the heat sources must be identified in phase 1 to ensure the positions of the sensors are optimally positioned for phase 2.

Procedure for testing:

1. With the battery pack positioned on the testing bench connected to the discharge machine, record and confirm the environmental testing temperature is at the specified requirements.
2. Photograph the battery pack from the top and side using the thermal camera to record the starting temperature.
3. Record the starting voltage of the battery pack to ensure it is at full charge.
4. Begin the discharge.
5. In intervals of 60 seconds, photograph the battery pack with the thermal camera.
6. Continue photographing at the set intervals until the battery pack either overheats or fully discharges.
7. Once either of the end conditions are met, photograph the pack for the after-discharge image.
8. Reporting. Reporting the maximum temperature reached and if the pack reached the cut off temperature.

Thermal Monitoring during discharge phase 2:

Description: Phase 2 can start once the location of the heat source has been identified, which are the hottest spots and the most representative of the battery pack. The locations can be used to position the Resistance Temperature Detector (RTD sensors), which can record the temperature of the identified location. The sensor locations are selected according to Globe group requirements for the battery pack design and based on other thermal tests conducted [31]. The locations selected for the temperature measurements with housing are shown in *Figure 11* and *Figure 12*.

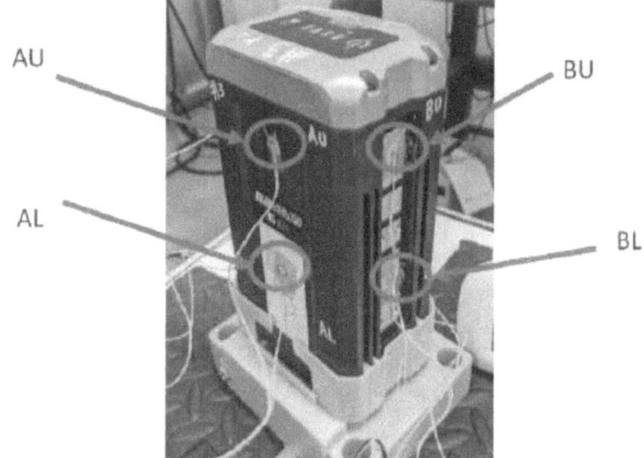

Figure 11: The positioning of sensors on the outer casing.

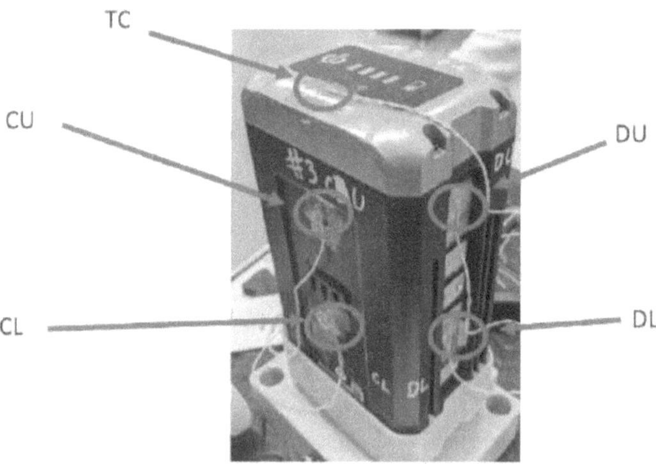

Figure 12: The positioning of sensors on the outer casing.

The locations selected for the temperature measurements without housing are in Figure 13*(a-b)*.

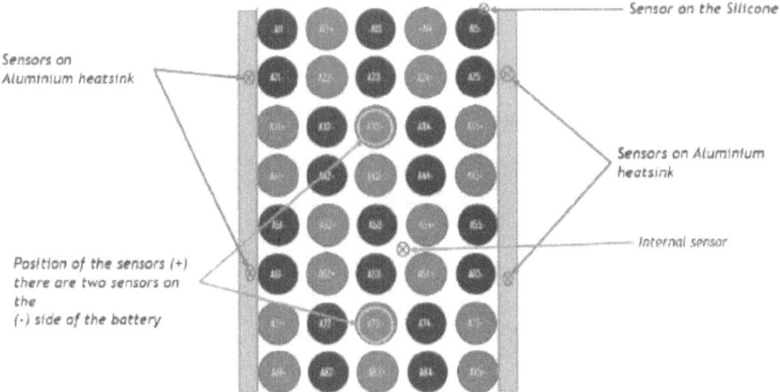

Figure 13: The positioning of sensors without outer casing.

Figure 13 a: The positioning of sensors without outer casing.

Figure 13 b: The positioning of sensors without outer casing.

During discharge tests, the internal battery temperature is also measured through the internal sensors with the help of a voltmeter, as shown in

Figure 14. Measure the internal temperature of the battery pack was essential for the testing, partly to see the correct temperature that the battery reads during discharge and partly to compare these temperatures with the other sensors around the battery pack.

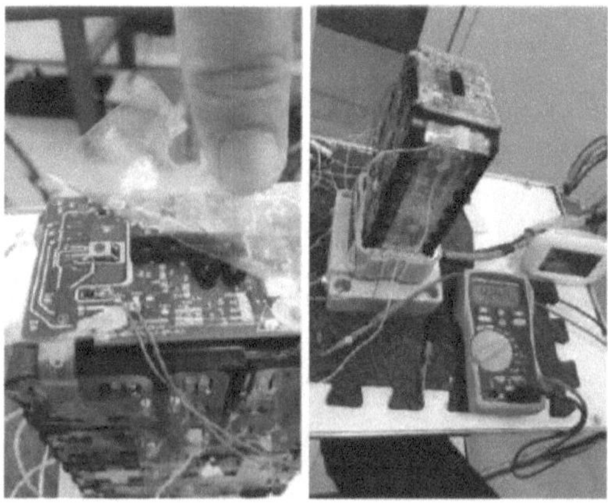

Figure 14: Measuring the internal sensor with a voltmeter.

The procedure of testing.

1. With the battery pack positioned on the testing bench connected to the discharge machine, record and confirm the environmental testing temperature is at the specified requirements.
2. Ensure the RTDs are securely mounted in the specified locations, and the recording device is measuring the temperature.
3. Report the starting voltage of the battery pack to ensure the battery is at full charge.
4. Connect the battery's internal sensors to the voltmeter.
5. Turn on the Unilogs to record the temperatures from sensors.
6. Begin the discharge.
7. Photograph with the thermal camera every minute throughout the test.
8. Discharge until the battery pack either overheats or fully discharges.
9. Once either of the end condition has been met, photograph the pack for the after-discharge image.
10. Allow the Unilogs to keep recording temperatures for another 10-15 minutes after the discharge machine stopped.
11. Turn off the Unilogs, END OF THE TEST.

3.3 Simulation Parameters

This section is a summary of how the parameters are entered into Solidworks for a thermal study.

3.3.1 Solidworks Materials:

The material selected for a component is based upon the manufacturing, aesthetic, and performance requirements of the component. In simulations, the performance of the materials is based on the material's properties such as density, which is why every component that is included in the simulations must have the appropriate material properties. For a thermal study in Solidworks, the only material properties needed are the material's density, thermal conductivity, and specific heat. These material properties significantly affect the results of the simulation and can vary greatly with different materials. It is crucial to identify the materials correctly and the properties it has.

To identify all the materials used in the battery pack, Greenworks was able to provide the full bill of materials (BOM). The BOM showed that the battery pack is full of many different components with different materials. The material properties of the components in the simulations must then be found and manually entered into the Solidworks material database. Materials in the real-life have a range for each material property depending on manufacturing technique or current environmental conditions. However, Solidworks requires the properties of the material only to use a single value.

3.3.2 Simulation Method:

Thermal simulation can be conducted with two different methods, either transient or steady state. A transient study starts the simulation from an initial temperature and runs for a set amount of time with interval time steps. Each time step provides a snapshot of the heat transfer occurring at that time; this progression of heat transfer is seen in Figure 15.

By increasing the number of time-steps, the simulation can provide more data points for the heat transfer. Increasing the number of time-steps also increases the real-life computing simulation time. The transient study also allows the temperature to be measured anywhere on the battery pack for any time-step, giving the ability to record data from the simulations, just like the real-life test. By being able to probe the same locations as the real-life test, the simulation results can be directly compared.

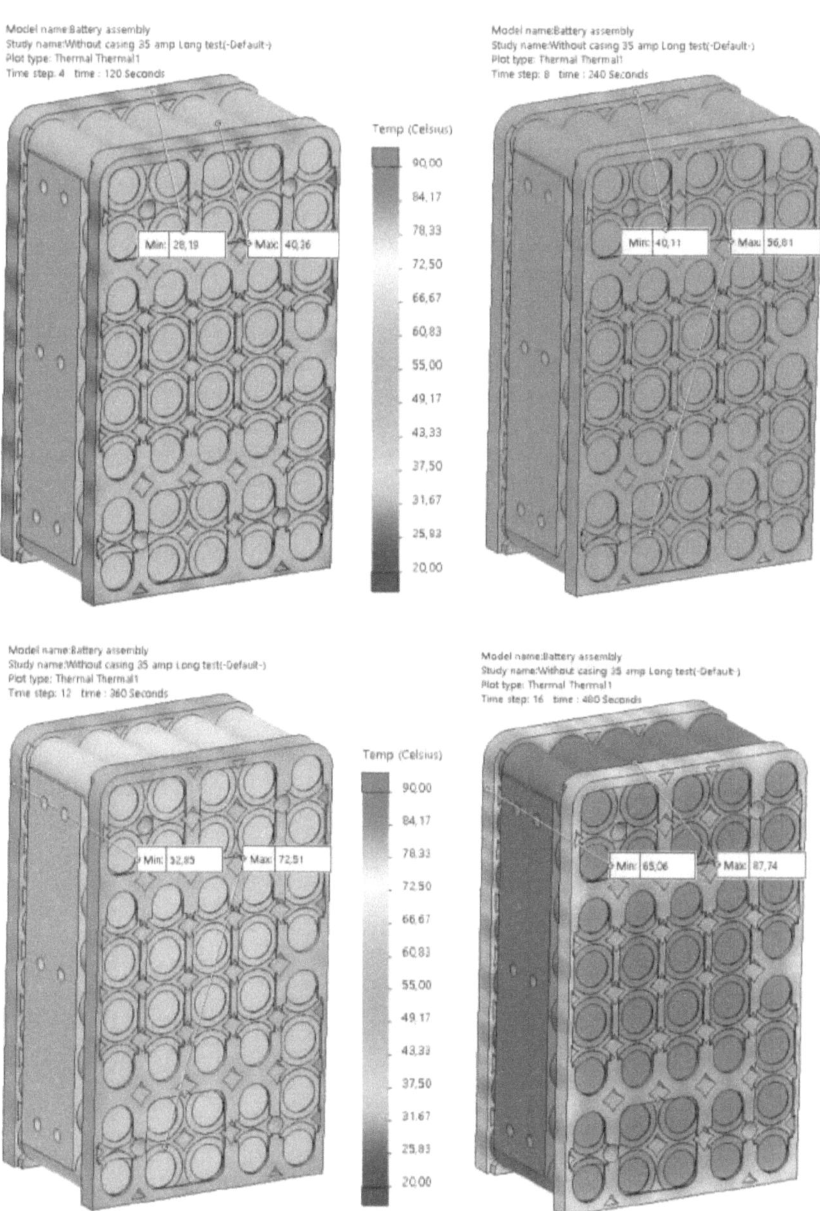

Figure 15: Distribution of heat at different time steps of the simulation.

3.3.3 **Simplifying CAD/Mesh:**

When a simulation model has many detailed components, the time to set up and run the simulation is drastically increased. Complex assemblies require more time to set up the relationship between each other, and more complicated relationships require more time to solve. Components with detailed features also increase the time by requiring smaller mesh designs, which increase the number of elements for the computer to solve. In the early stages of simulation, a good practice to save time is to reduce the number of components in the assembly and simplify the complexity of each part.

All parts in the simulation should be considered, but in the early stages of design and development, smaller or nonrelevant parts can be excluded to save time. Removing these components does not affect the result as much since they only contribute a small percentage of the overall system. This is the same for complicated, detailed parts, removing these small features such as fillets, decorative items, or nonrelevant features allows the mesh to be larger and faster to solve. The complexity of a model is also relative to the computer resources available; faster computers will have no issue solving complex geometry but simplifying the model will always save simulation time. Once the simulation parameters and design of the product is finalized, then a detailed simulation is appropriate to use.

The simplification of the model played a significant role in meshing the model. The original model provided was required a very small mesh requiring less than 0.5 mm for the maximum element size. After simplifying the model, the maximum size of the mesh elements got up to 4.5 mm while any larger caused the mesh to fail and simplifying the model anymore would not correctly represent the real-life battery pack. The mesh type used was the standard Solidworks mesh known as the Voronoi-Delaunay meshing scheme, as seen in Figure 16. The other meshing options caused failures and required a smaller element size. During the testing and evaluation of the simulations reducing the simulation time was always considered to be able to performer more simulations in a day.

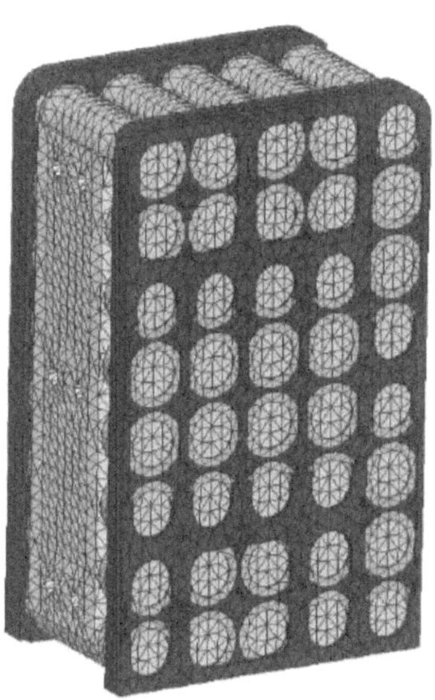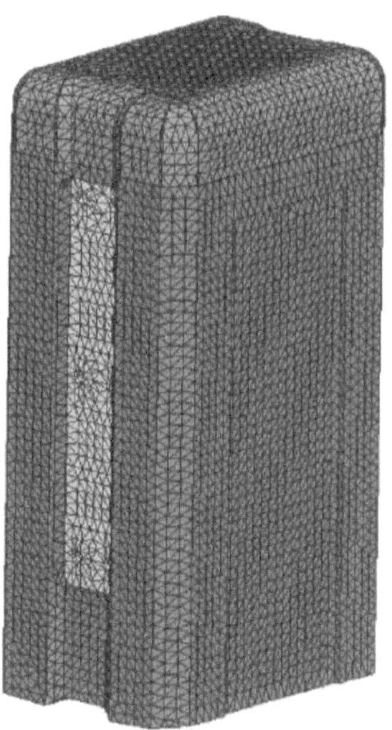

Figure 16: meshed of the simplified battery pack with & without casing.

3.3.4 **Initial temperature:**

For a transient thermal study, there must be an initial temperature for all the components in the systems. Just like in real life, everything has a temperature, and for transient studies, the temperature must be known before the thermal study starts.

3.3.5 **Conduction and Thermal Resistance:**

The main spread of heat throughout the system is due to the physical contact between parts, also known as conduction. The change in temperature due to conduction is determined by the thermal resistance between each part in contact. The thermal resistance is determined by the material's thermal conductivity, the part thickness, and area of surfaces in contact between the parts in question, as seen in *equation (9)*. The surfaces in contact with each other must be identified and sectioned off to apply the thermal resistance in Solidworks appropriately. The value for the material thickness is measured in the CAD model for the specific locations. The value for the thermal resistance is hand calculated and entered Solidworks via contact sets. Fortunately, for the surfaces that have complicated surface areas to calculate, Solidworks can apply a distributed thermal resistance where only the thickness and thermal conductivity is required.

3.3.6 **Convection:**

To represent the effects of convection heat transfer in Solidworks, each surface that is exposed to the air must be selected to apply this effect. This type of heat transfer is described in *equation (15)* where h is the Heat transfer coefficient, A is the surface area, Ts is the surface temperature and, T_f is the temperature of the fluid. Solidworks can account for the surface area, and temperatures but requires the convection coefficient value from the user. Solidworks help page provides an estimation of what the convection values can be for different cases seen in *Table 7.*

Medium	Heat Transfer Coefficient h ($W/m^2 . K$)
Air (natural convection)	5-25
Air/superheated steam (forced convection)	20-300
Oil (forced convection)	60-1800
Water (forced convection)	300-6000
Water (boiling)	3000-60,000
Steam (condensing)	6000-120,000

Table 7: Typical values for the convective heat transfer coefficient.

3.3.7 **Radiation:**

Radiation is the most complicated type of heat transfer for Solidworks to solve. In real-life, every object above 0 degrees Kelvin is emitting and absorbing thermal radiation in every direction. Solidworks simulates this effect, in the same way, using a method called view factor, or configuration factor where it accounts for direction and distance of every surface in the model as a path for the radiation partials to travel. Solidworks must also calculate how each surface that is emitting radiation interacts with each other in terms of heat transfer. Solidworks having to account for both in each time-step significantly increases in simulation time. When assigning the parameters for larger simulations, radiation should also be considered in terms of practicality.

3.3.8 **Heat Power:**

Heat power is used to represent the heat generated in the Solidworks simulation. The heat power can be applied to surfaces and bodies in the model using watts as the unit of heat energy.

3.4 Comparison between simulation and experiment

The experiments conducted in the lab are used as the baseline for the simulations. The idea is to match the simulation results with the real-world tests; once these values match each other, then new designs can be tested. The purpose of comparing the simulation and the real-life test is to make the testing conditions for the simulation and the real-world the same. If the testing conditions are the same, then new designs can be tested in the simulation without manufacturing anything in the real-world.

The procedure for the comparison is to simply collect temperature data from the same location on both the simulation battery pack 3D-model and real-world battery pack under the same conditions. The collected data is compared against each other and determined if the simulation parameters are acceptable conditions for testing other designs.

As mentioned before, the simulations largely depend upon the input parameters from the user, such as thermal conductivity, convection of air, starting temperature of the test system, and much more. The issue and benefit of the simulation are seen here, while the simulation will "run the test" under the conditions and parameters specified, it can run it the test perfectly under the conditions specified. The problem with this is that the real world is never perfect, and there can be differences between the simulation and the real-world test. These differences occur in many ways, for instance, the value for Aluminum 6063-T5 thermal conductivity can be between 201-217 W/mK, and the simulation parameters only allows one value to be selected. In addition, the aluminum in the real-world test may have very different values than the material properties listed. These differences between the real-world and simulations are where most work is conducted for this book.

4 Results and analysis

This chapter presents and discusses the result from experiments and simulations. The simulation parameters used in the final simulations are explained.

4.1 Experimental results

Result from thermal Monitoring during discharge phase 1:

The results of Thermal monitoring during discharge phase 1 shows how the temperature increases and spreads out during discharge. The data from the Thermal Camera did not show any hotspot in the battery pack and showed that the Heat increase quite homogeneous overall during the discharge test of with/without housing of the battery pack, as it is shown in Figure 17.

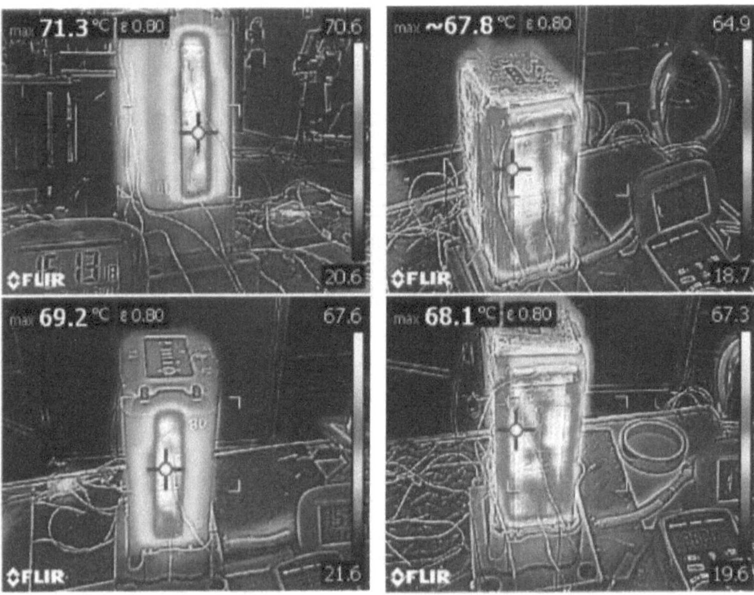

Figure 17: Infrared pictures from the thermal camera.

As *Figure 17* shows, there is a significant temperature difference between the outer housing and the heatsink.

The PCB is investigated with a thermal camera during testing to determine if the PCB or are any components were causing the overheating issue, from Figure 18 the thermal camera does not show any hot spots on the PCB.

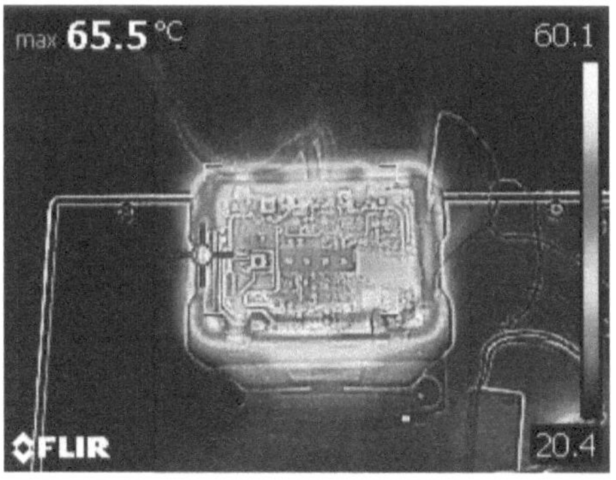

Figure 18: Infrared picture of the PCB during the discharge.

Result from thermal Monitoring during discharge phase 2:

As described in the experimental section, the batteries were tested at different discharge rates; these tests are listed in *Table 8*.

List of test	12-feb	17-feb	19-feb	21-feb	24-feb	25-feb	26-feb
Battery # 1	30A W/ Casing - 77A internal temp test	30A W/out Casing	35A W/out casing - 35A W/ casing			55 A W/ Casing	
Battery # 2		30A W/ Casing					
Battery # 3	Batteries have not arrived			35A W/ Casing	40A W/ Casing	45A W/ Casing	45A W/out Casing
Battery # 4				35A W/ Casing	40A W/ Casing	45A W/ Casing	
Battery # 5				35A W/ Casing	40A W/ Casing	45A W/ Casing	Non-CC 55A15sJA20s
Battery # 6				35A W/ Casing	40A W/ Casing	45A W/ Casing	Non-CC 55A25s0A60s
Battery # 7				35A W/ Casing	40A W/ Casing	45A W/ Casing	45A W/ Casing Al Test

List of test	05-mar	12-mar	16-mar	17-mar	02-apr
Battery # 1	35 Amp W/casing (Thermal paste)				
Battery # 2					
Battery # 3	45 Amp sensors inside casing	35 Amp without casing	45 Amp without casing		35 Amp with casing
Battery # 4		35 Amp without casing	45 Amp without casing		35 Amp with casing
Battery # 5	35 Amp W/casing		35 Amp without casing	45 Amp without Casing	35 Amp with casing
Battery # 6			35 Amp without casing	DEAD	
Battery # 7		35 Amp without casing	45 Amp without casing		35 Amp with casing

Table 8: List of conducted tests.

The first tests that performed on batteries were with casing on, due to keeping the sealing of the batteries for a better testing result. During testing, it was discovered that batteries 1 and 2 have different configurations, because of this, none of the results from these two batteries are included in the result section. When the two batteries were examined, it was discovered that they had two additional heat sinks that batteries 3-7 did not have. Results from discharging at 35 and 45 Amps show in *Figure 19* and *Figure 20*.

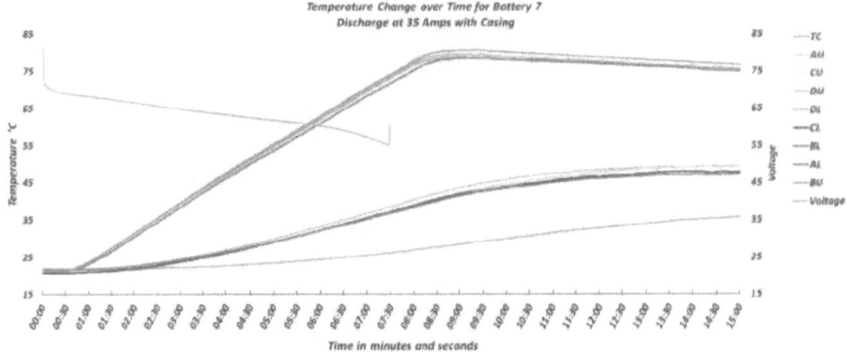

Figure 19:Discharge at 35 Amps with casing - battery 7.

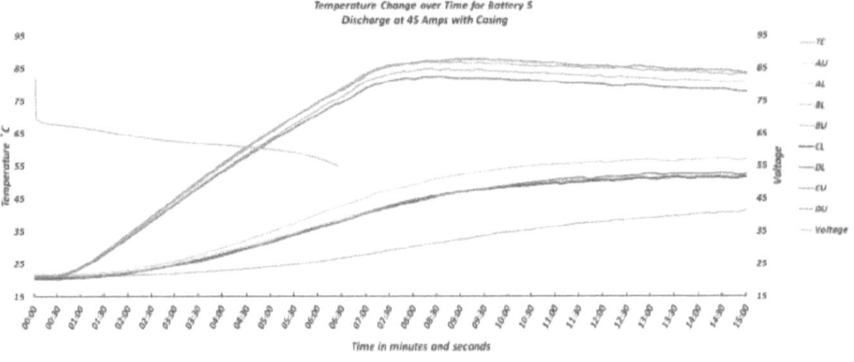

Figure 20: Discharge at 45 Amps with casing - battery 5.

Non-continuously discharge at 55 Amps with an interval of 25 seconds discharge and 60 seconds pause, in this test, investigated the heat transfer of the battery under non-continuous discharge with high current. The result shows how the temperature stabilizes and rises again during discharging at the heatsinks. A closer look at the temperature plot shows that the temperature rise delays a little after each discharge interval due to the time required for the temperature to rise in the battery cell and spread until reaching the heatsink. The result is shown in *Figure 21*.

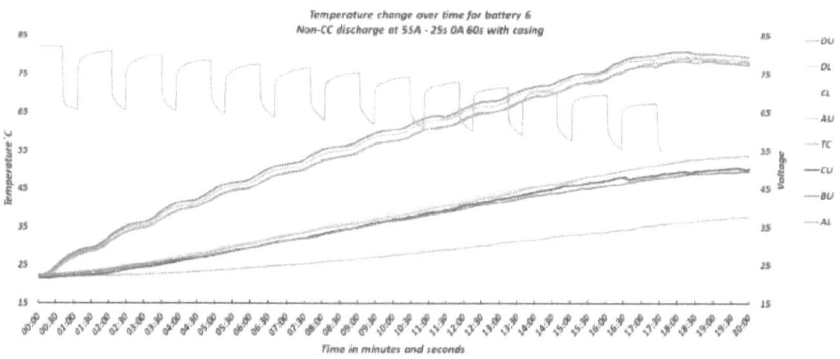

Figure 21: Non-CC discharge at 55Amps (25s0A60s) with casing battery 6.

The temperature and positions for the battery packs tested without outer housing are shown in *Figure 13*. The temperature of internal sensors was also measured during the discharge tests and is shown in Figure 22.

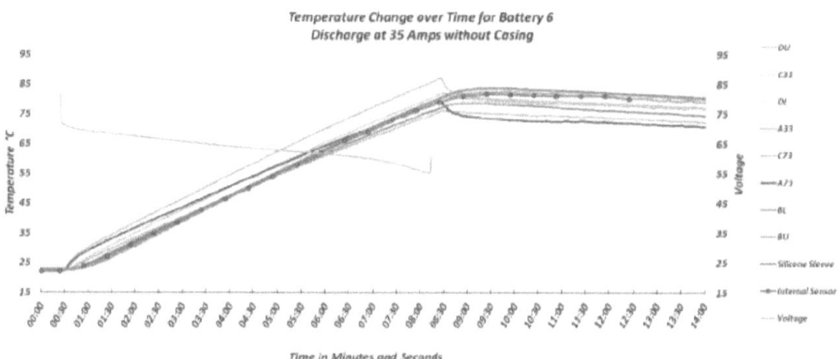

Figure 22: Discharge at 35 Amps without casing-battery 6

The batteries tested at 45 Amps rapidly increased their temperatures to 75 C before the battery pack was able to reach 55 V. For the safety of the test, these battery packs were shut down earlier than the usual 55V cut-off. The results are shown in *Figure 23* and *Figure 24*.

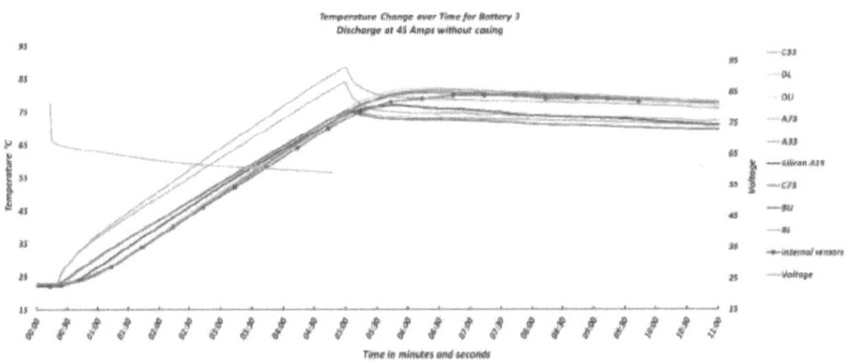

Figure 23: Discharge at 45 Amps without casing – Battery 3

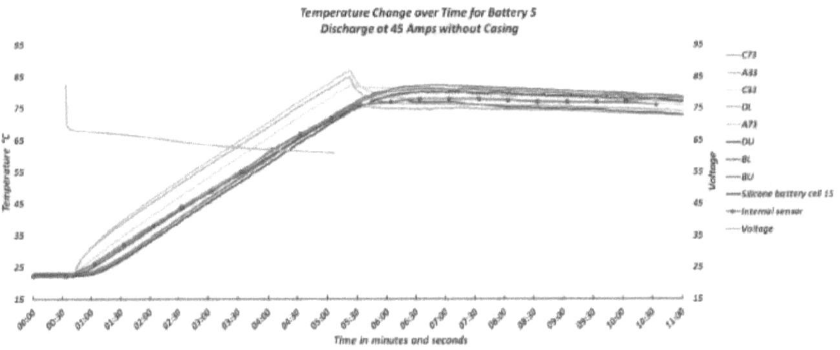

Figure 24: Discharge at 45 Amps without casing – Battery 5

The collected data from discharging tests have been compared with similar positions in other batteries, to see if all batteries behavior similar respect to each other. The comparison results of few positions on different discharge rates shown in *Figure 25(a-c)*.

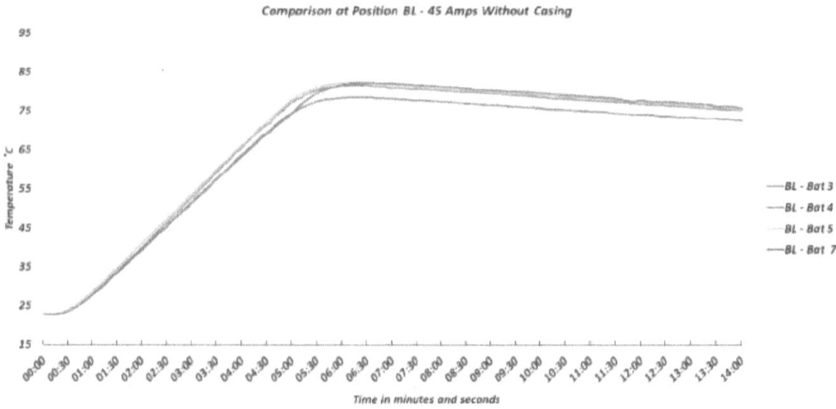

Figure 25: Comparison of silicone position at 45 Amps without casing.

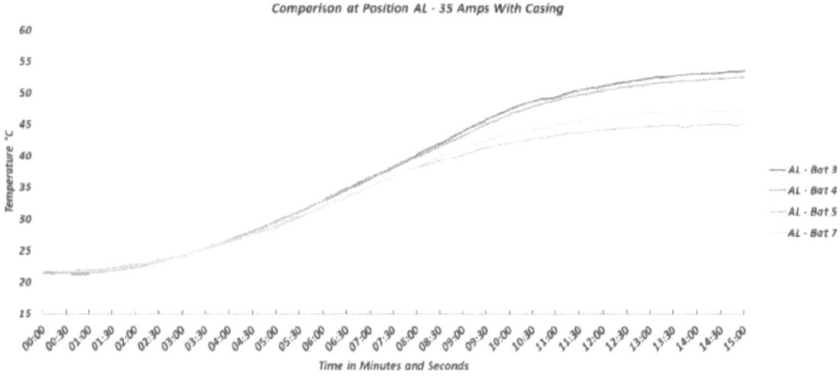

Figure 25 a: Comparison of position AL at 35 Amps with casing.

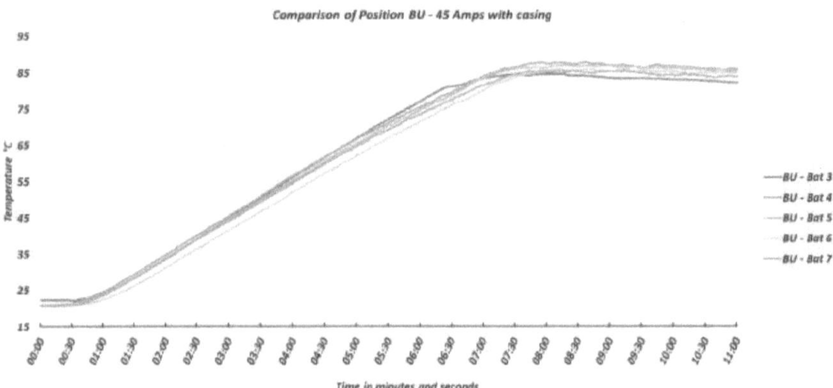

Figure 25 b: Comparison of position BU at 45 Amps with casing.

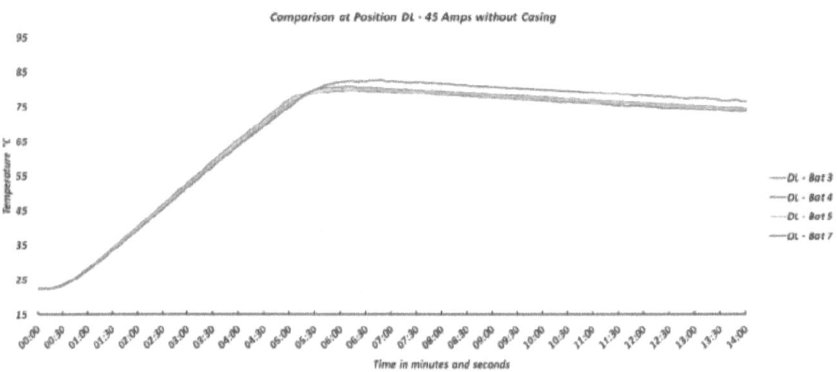

Figure 25 c: Comparison of position DL at 45 Amps without casing.

Figure 26 (a-c) displays the comparison of the sensors on the ends of the battery cells. What is quite impressive is that positions C33 & C73 are on the negative side of the battery and behave similarly to other positions in the study, but positions A33 & A73 on the positive side behave very differently. It seems the positive side of the battery is rapidly heating and cooling due to the direction the electrons are move between the batteries.

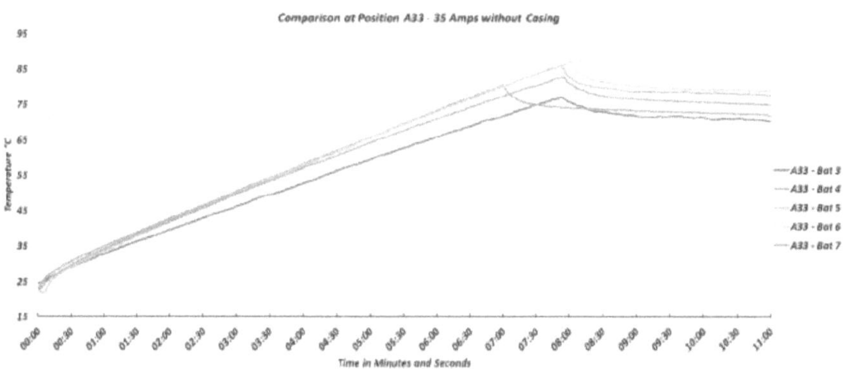

Figure 26: Comparison of position A33 at 35 Amps without casing.

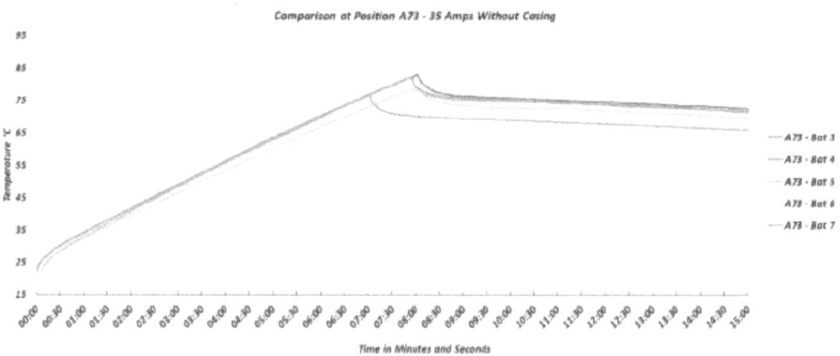

Figure 26 a: Comparison of position A73 at 35 Amps without casing.

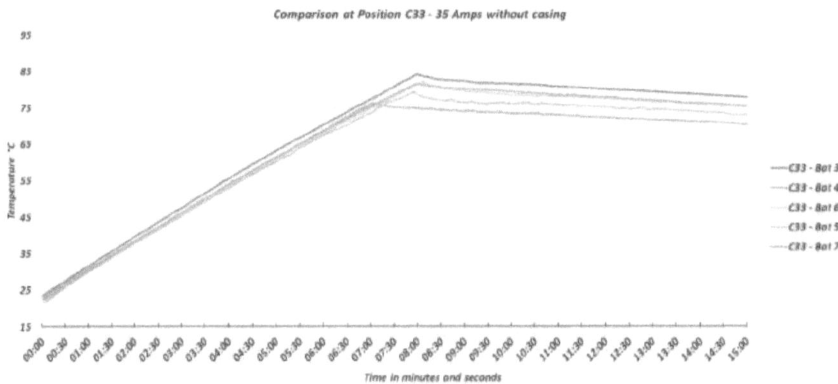

Figure 26 b: Comparison of position C33 at 35 Amps without casing.

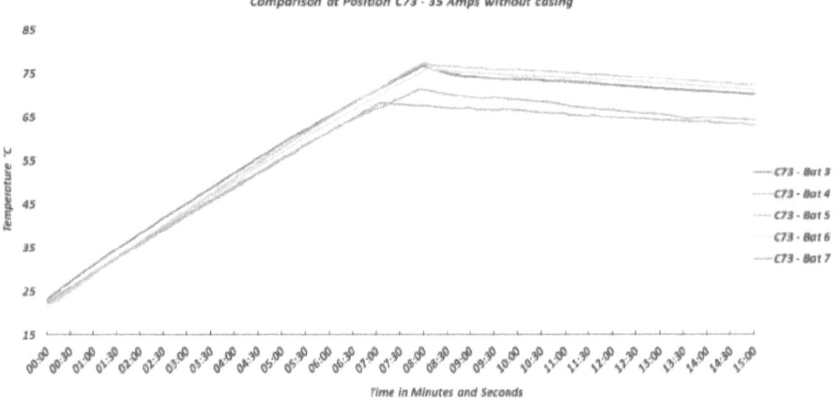

Figure 26 c: Comparison of position C73 at 35 Amps without casing.

4.2 Simulation Parameters

This section reviews the parameters used in the simulations, and how the values for the parameters were selected. This section includes how the source of heat was modeled, all the three forms of heat transfer in the system, and the alternative method used to create the simulations.

4.2.1 **Battery Pack Materials**

For the thermal simulation, only the more significant components or components that have a direct influence of the thermal study are considered. All the components inside of the battery pack influence the thermal behavior; the components that are excluded from the study are much smaller relative to the system and influence the thermal behavior much less. These smaller components are excluded from the study to reduce the study complexity and simulation run time. The materials for the components used in the battery pack are lists in the BOM. The material properties of each of the materials were found in the database "Material Universe," also known as "CES." The lithium-ion battery cells and the silicone battery sleeves were the only two parts that could not be found in the CES database.

The Cramer BOM specifies that the silicone sleeved had a thermal conductivity of at least 2 *W/m.C*; the other of the properties had to be estimated with CES. Lithium-ion batteries were researched and found to have a wide range of thermal properties based on the cell construction and materials used in them, the values used for the simulation are based on values found from U.S. Naval Research Laboratory [32]. The compiled list of components and materials properties used in the Solidworks simulations are listed in *Table 9.*

Part	Aluminum	Plastic Holder	Battery Cell Sleeves	18650 Cells	Outer Casing	Top Cover
Material	6063 T5	PA66-GF30	Silicone		ABS+PC (Flame Retardant)	Magnesiom Alloy AM60B
From Research and CES						
Thermal Conductively W/m.C	201-217	0.4-0.55	2	3.1-3.6	0.276-0.287	60-64
Density kg/m3	2660-2710	1510-1540	2150-2300	2725	1170-1230	1770-1790
Specific Heat J/kg C	881-917	1480-1530	1050-1100	900-1020	1530-1600	1030-1070
For Simulations						
Thermal Conductively W/m.C	209	0,5	2	3,35	0,282	62
Density kg/m3	2700	1525	2225	2725	1200	1780
Specific Heat J/kg C	900	1500	1075	960	1565	1050

Table 9: List of components and materials properties.

4.2.2 **Simulation Method**

For the 35 Amps test, the total simulation time is 660 seconds or 11 minutes with time steps every 30 seconds. During the real-life battery test, the batteries took approximately 8 minutes & 30 seconds to discharge the battery pack to the cutoff voltage of 55 V. The batteries were then monitored for an additional 10 minutes to observe the convection cooling rate. From analyzing the temperature curve after the discharge machine cut off, it was determined that the battery is cooling at a linear rate. This linear decrease in temperature meant there would be no value in continuing to simulate for more than 11 minutes. This allowed the simulation to collect 5 more data points after the heat power shut off to ensure the simulation was cooing at the same linear rate. This same practice for the 45 Amps simulation where the real-life 45 Amps discharge test lasted for approximately 5 minutes and the simulation ran for 8 minutes 20 seconds.

4.2.3 **Simplifying CAD/Mesh**

The CAD model from Globe Group had detailed part geometry and a complete assembly of the battery pack. While this realistic model, seen in *Figure 27*, is useful for visualizing the design and testing manufacturability, the model is challenging to simulate. The only components included in the simulation are the Li-ion battery cells, the aluminum heat sinks, the silicone sleeves the batteries sit inside and the battery holders. All other components are excluded, to simplify the assembly as these components were viewed as unessential to the thermal model or would be poor representations of the real-life component. The simplified CAD assembly, seen in *Figure 28,* requires much less effort for the setup and decreases the simulation run time without losing much in the accuracy of the results. In addition to reducing the number of parts in the assembly, the remaining components were then simplified even more by removing the fillets, ornamental features, and complex surfaces. Removing these smaller detailed features allows the simulation mesh to be larger, which contributed to decreasing the simulation time. By removing the fillets this will decrease the heat flux in the corners causing the temperature to be greater than in reality. This trade off in accuracy of the model was acceptable to decreases the simulation time.

Figure 27: CAD model of the Cramer 82V battery pack.

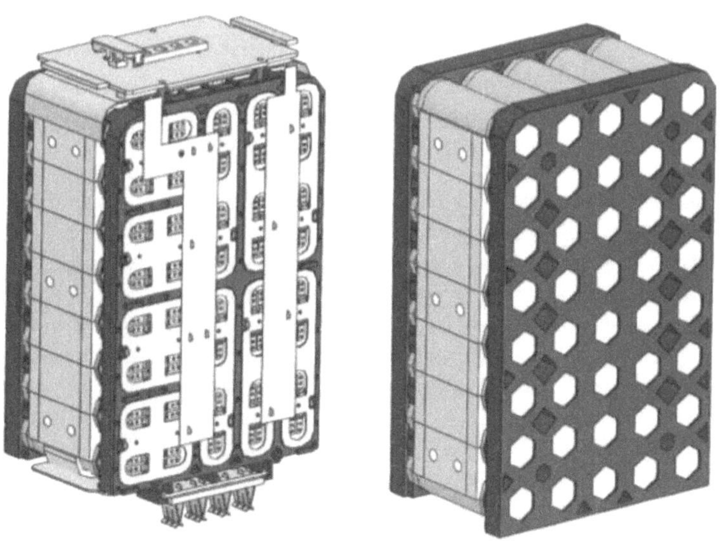

Figure 28: Original and Simplified CAD model.

4.2.4 **Initial temperature:**

The initial temperature for all the bodies in the simulation was 22-degrees C. This value is from the component's temperature before starting the tests.

4.2.5 **Conduction and Thermal Resistance:**

In most cases applying the thermal resistance, the contact set posed no issue given the correct thermal conductivity values and thickness identified. The only issue that occurred with determining the thermal resistance value is for the Lithium-ion cells. The 18650-battery cells have complicated construction, which leads to the difficultly in knowing what material properties to use the represent the battery. As shown in *Figure 1*, the lithium-ion battery cells are made up of many layers of alternating materials, but all these materials are housed inside a stainless steel can with a layer of insulating PVC vinyl. Considering that thermal resistance is calculating the contact resistance between the outer contacting surfaces, the thermal conductivity coefficients values for the two outer layers of 18650 cells, the can, and the insulator, were used for the lithium-ion battery.

4.2.6 **Convection:**

The value for the convection coefficient provided by Solidworks in Table 7: Typical values for the convective heat transfer coefficient., did not match the data from the real-life test.

The convection coefficient was tested by observing how the battery cooled down after the heat power shut off. When using the Solidworks values (5-25 *W/m2.K*), the entire battery would cool down much faster than what was recorded in the real-life test. Using the correct value for the convection coefficient is vital since the convection heat transfer is always active during heat power and cooling phases of the simulation.

Further research for the convection coefficient determined that the value could vary greatly depending on the change in surface temperature and the geometry of the surface seen in *equation (20)* and *Table 12* in the appendix. The geometry of the battery pack surfaces was assumed to be either horizontal or vertical plates. The equations for both types of geometry are used to solve for these convection coefficient values at every temperature between 22-150 degrees Celsius. A MATLAB code was written and used to calculated and plot the coefficient values seen in *Graph 1*. Both curves were tested in Solidworks simulation but found to be too high still.

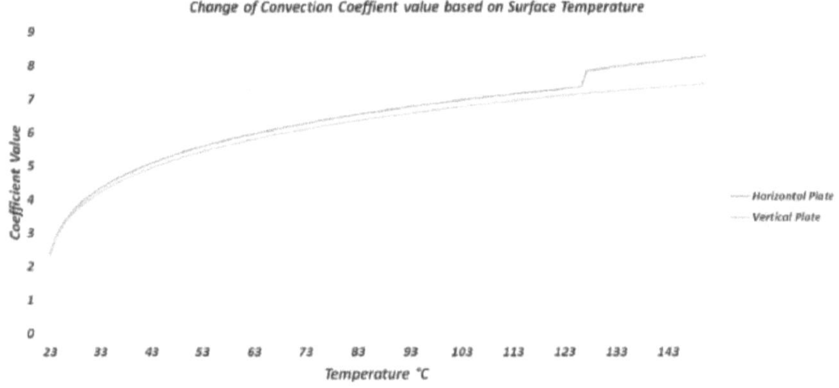

Graph 1: Convection coefficient curves from MATLAB.

Along with the MATLAB calculations, [30] was used to determining the convection coefficient value. This article was from heating, ventilation, and air conditioning (HVAC) investigated, which studied the change in convection coefficient values as the temperature in the room changed. This study found that depending on the room conditions and heat from the windows, the convection coefficient values could be modeled with equations in *Table 10*.

Room Window	Equations
No Blinds	$h = 1.74x^{0.33}$
Window: -45 Degree Blinds	$h = 1.89x^{0.25}$
Window: 0 Degree Blinds	$h = 1.67x^{0.25}$
Window: +45 Degree Blinds	$h = 1.48x^{0.25}$

Table 10: Impact of blinds on convection along with windowpane.

Both the MATLAB and HVAC room studies can be plotted in *Graph 2*. Testing each of the convection coefficient curves seen in the graph below, the curve for Windows +45 degrees (green line) provided the best result when compared to the real-life battery testing.

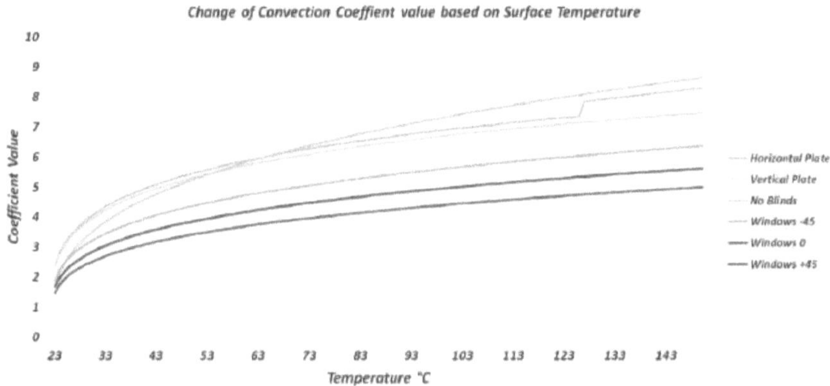

Graph 2: convection coefficient curves in various room conditions.

4.2.7 Radiation:

Based on *equations (21)* and *(22)* and the time requirements for the simulations to solve, radiation is not considered for the final simulation design. The addition of radiation in the simulation, increased the computational time by 3-12 hours for each time step, and only accounted for approximately 8 Watts, just 2% of total power in the system.

The low radiation power is due to two factors, the temperature of the battery pack and the surface emissivity. Compared to the ambient air, the temperature of the battery pack is relatively not so significant. The other major factor is the surface emissivity of the battery pack. In *Figure 13 a* seen that most of the pack is covered with the reflective nickel and aluminum surfaces, which according to *Table 13* for material emissivity, the values are quite low.

The radiation heat transfer is not in the simulation parameters because of the low impact and the time saving for the simulation,

4.2.8 **Heat Power:**

From the beginning of the study, only ohmic heating was used to calculate the heat power produced from the battery cells. This method used *equation (8)* and only required two factors the batteries internal resistance (impedance) seen in *Table 5* and the current flowing through the battery cell. The specification sheet for the Lithium-ion batteries used in the battery pack has an impedance ranged between *8-18 mOhms* per cell. According to [33], This range of impedance has an inverse relationship with the battery cell temperature, where the electrical resistance decreases as the temperature increased. When calculating the heat power, the highest value of 18 *mOhms* was used as the internal electrical resistance, and this was assumed because the initial temperature was on the lower end of the battery cell operating temperature. While according to [33], the impedance should decrease with the temperature, the rate at which this would occur is unknown and required more in-depth research into the individual battery cells themselves.

The current passing through the battery cell is based on the load applied by the discharge machine and the cell configuration. As seen in *Figure 29,* the cells are arranged in a 2S20P configuration with connecting bridges in between each parallel set, *Figure 30* and *Figure 31* shown the arrangement of the cells and how they are connected to each other inside the battery pack. According to *ohms law's*, this arrangement would split the load between the two cells that are in parallel with each other. Assuming that all the cells in the battery pack have the same internal electrical resistance, the total load on the battery pack is half for each cell.

The value tested and calculated was 5.5125 W/cell, the simulation found this value to be far too low, further [16] suggested that ohmic heat for Lithium-ion batteries only account for 50%-58% of the total heat generated by the battery cell. The portion of the uncounted heat power comes from the change in entropy and the exothermic chemical reaction occurring inside the cells.

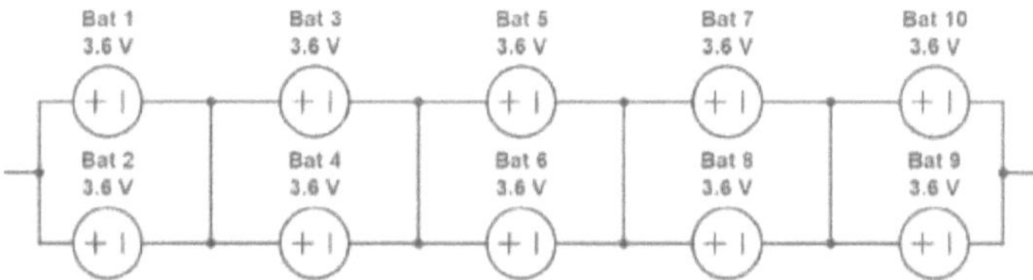

Figure 29: Simplified example of the cell configuration.

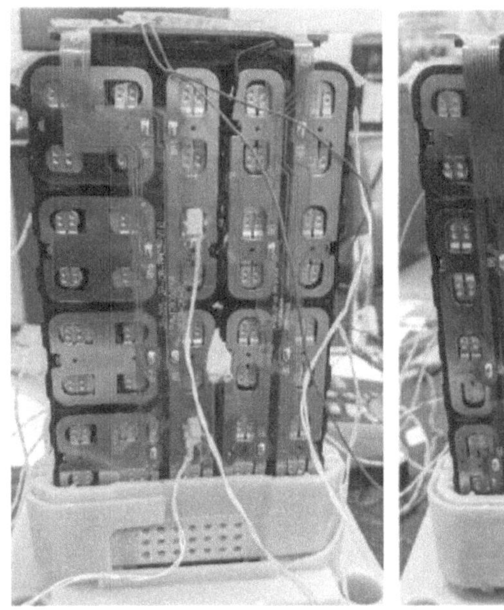 

Figure 30: Arrangement of the cells in the battery pack.

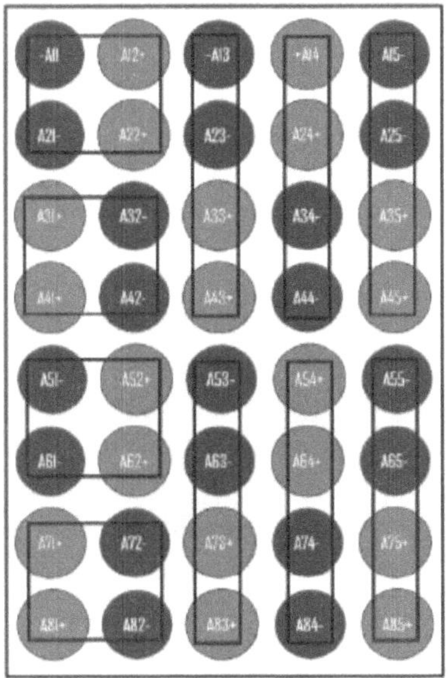 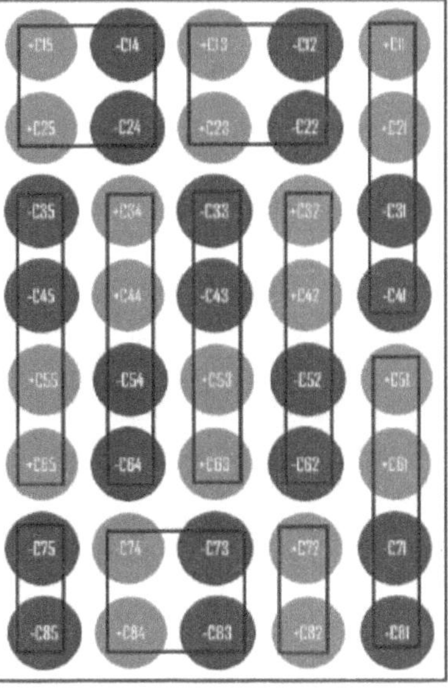

Figure 31: Simplified image of how the Li-ion cells are welded to each other.

4.2.9 **Additional Heat Power Calculations**

Since the preliminary heat calculation could not provide a heat power value that matches what was occurring in discharge tests, two alternative methods were used to find the final heat power value. The first step was to double-check the amount of heat energy required to heat the battery pack to temperatures it was reaching. By measuring the battery pack's thermal capacity shown in *equation (24),* which only requires the mass, specific heat values, and the change in temperature over time, all of which were available after the discharge tests. The MATLAB code in appendix 7.6, is used to calculate the minimum amount of heat power that is required for materials inside the battery pack to reach the temperatures that were measured in the experiments. *Graph 3* shows the calculated values from MATLAB, which averaged out to be 8.08 Watts of heat power. The graph is the sum of heat power required to heat the materials inside the battery per second.

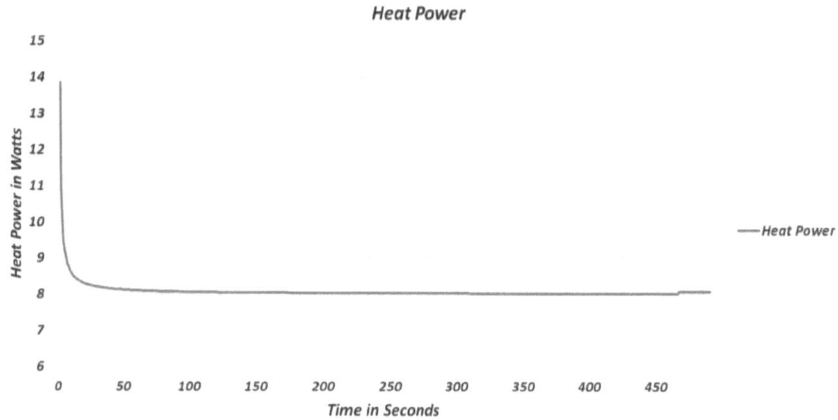

Graph 3: Minimum required heat power calculated by MATLAB.

The heat capacity calculation found that it would require at least 8.08 Watts of heat power to heat the materials in the battery pack. This value confirmed that the initial value that was calculated was far too low. *Figure 32* shows the initial heat power value of 5.5125 watts with the blue line is far below the test results; in green, is the new value calculated based on the heat capacity.

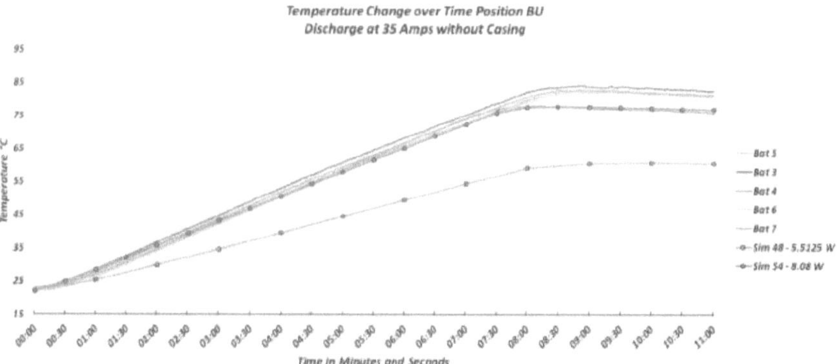

Figure 32: Comparison of initial heat power value (sim 48) and heat power value from MATLAB (sim 54).

The 8.08 W value is still a bit low, but as it was calculated, this value should the minimum heat power. The next step was to optimize the heat power value based on the experimental results. Solidworks can optimize the heat power but running multiple simulations back-to-back, slightly altering the heat power in each simulation. The objective is to match a targeted temperature at a specific time and location. The selected target for the optimization was the 70°C at position BU, and 375 seconds in the simulation, this was the highest temperature measured at position BU at this time. The new value for the heat power from the optimization is 8.6 watts, this value for heat power is slight on the high end but matches the experiment result quite well, as seen in *Figure 37(a-g)*.

4.2.10 **Modeling Air**

The simulation with the casing posed another issue, to correctly transfer the heat from the battery cells to the outer walls of the protective casing. The issue was solved using a part to represent the air between batteries and the casing. This part was based on the theory of *equation (17),* which stated that if the *Nu* number is equal to 1, the fluid transfers heat like a solid. This theory was assumed and applied because the casing had an ingress rating of *IPX4,* so air should have very little flow and should stay inside the casing. *Figure 33* is the part that represents the air; this part has the density and specific heat of air based on *Table 14.* The value for thermal conductivity was treated as the convection coefficient since there is a finite amount of air inside the battery pack, and it cannot escape. In real-life, the air would continually spread the heat into the ambient air allowing cooler air to take its place. In this case, the air continues to heat up without an outlet; this effect is represented by using the air's thermal conductivity value as the convection coefficient.

The value for air's thermal conductivity was determined by optimizing the temperature at position AL on the outer casing. The optimization determined that the thermal conductivity value should be 20 W/m.K. This value is in line with the suggested values for the convection coefficient seen in *Table 3* and *Table 7*.

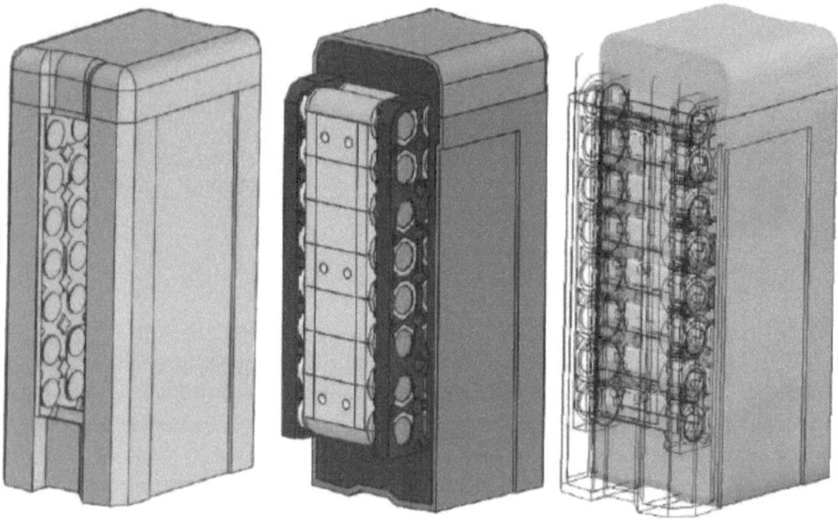

Figure 33: Right: cross-section view of the battery pack, Middle: cross-section of the simplified battery pack and Left: 3D model to represent the air.

The left displays the simplified modeled of the battery pack with the outer casing. The part on the right is the model of the air. This part is inside the battery pack and fills all the gaps between the batteries and outer walls.

4.3 Simulation Results

The results from the simulation show that the heat is evenly distributed throughout the battery pack, as shown in *Figure 34* and *Figure 35*. These figures suggest that all the thermal conduction contacts and the convection values are correctly assigned. The model of heat in the simulation is suitable for displaying how the heat spreads out and what areas are warmer and colder, but the interesting data is in the graphs.

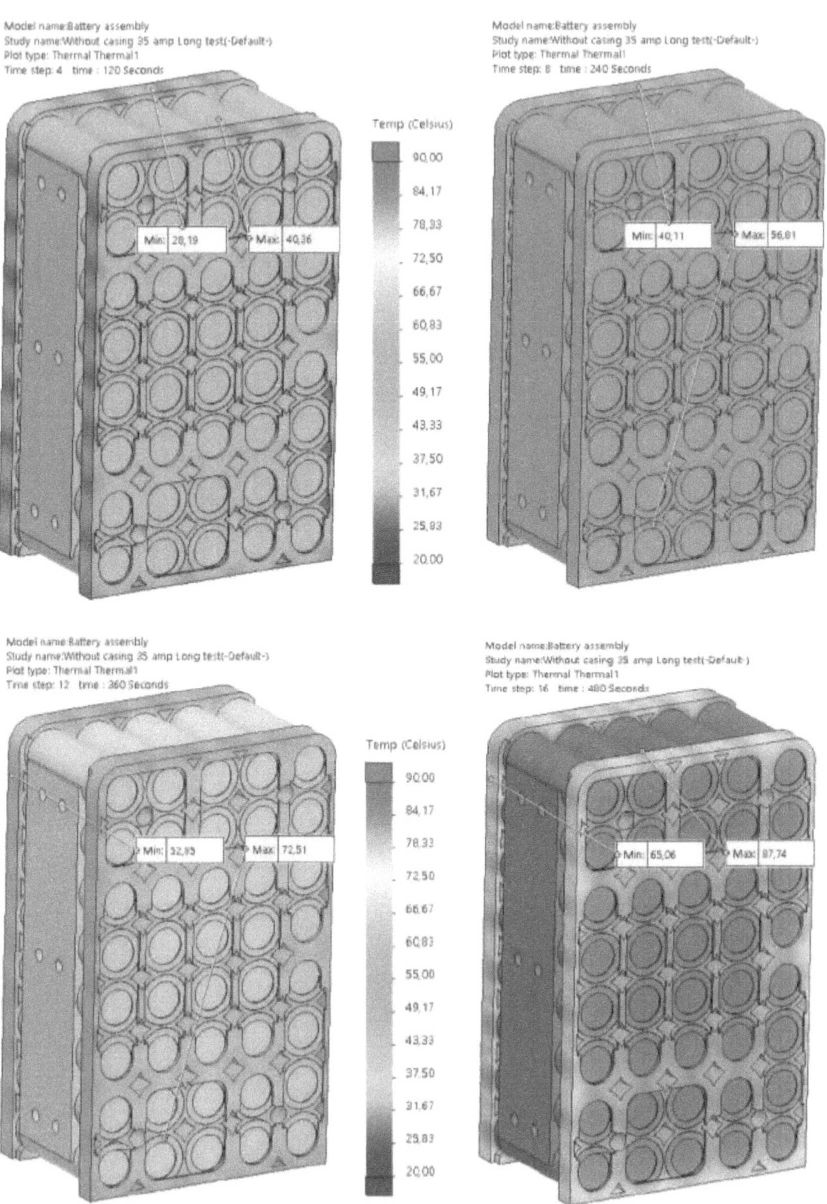

Figure 34: heat distribution in different steps of simulation at 35 Amps.

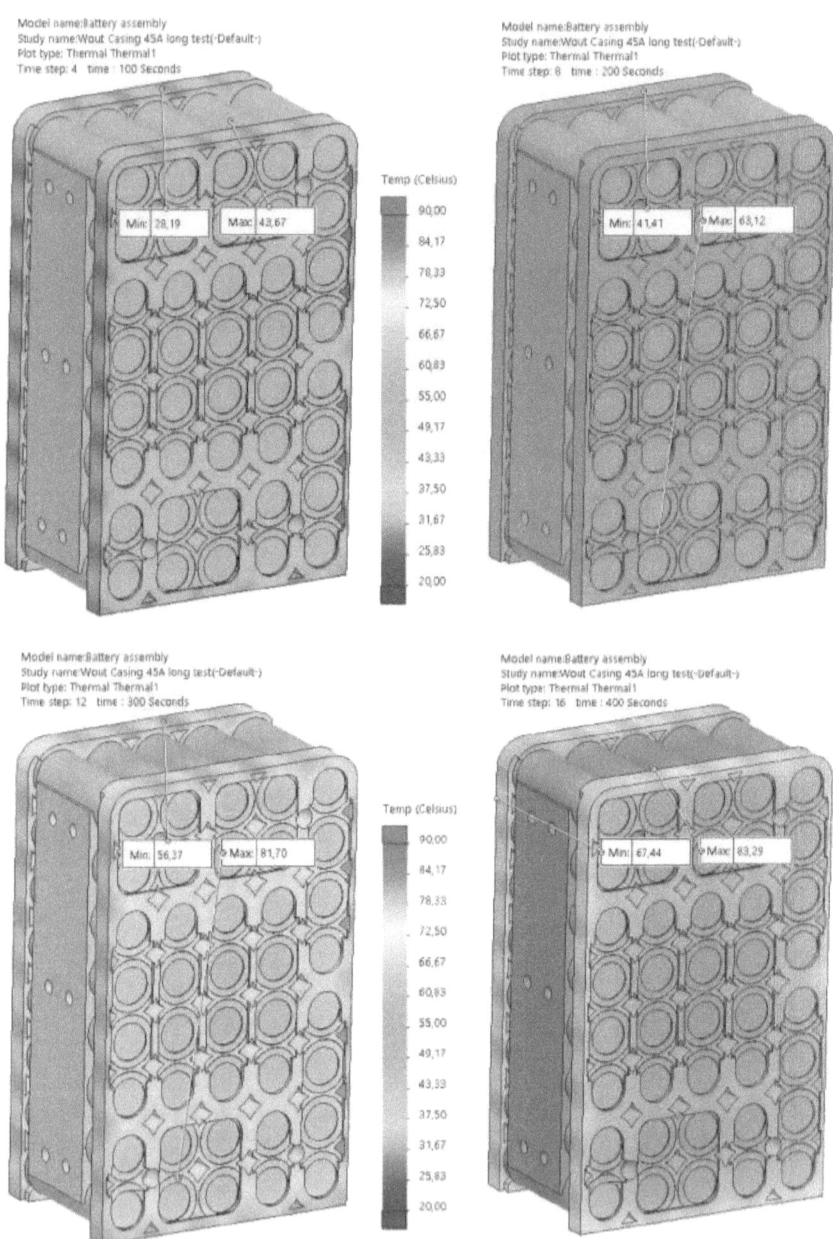

Figure 35: heat distribution in different steps of simulation at 45 Amps.

Figure 36 shows how the heat takes a long time to spread from the center of the battery pack to the outer walls. The outer walls of the battery pack continue to heat up long after the battery heat power is shut off.

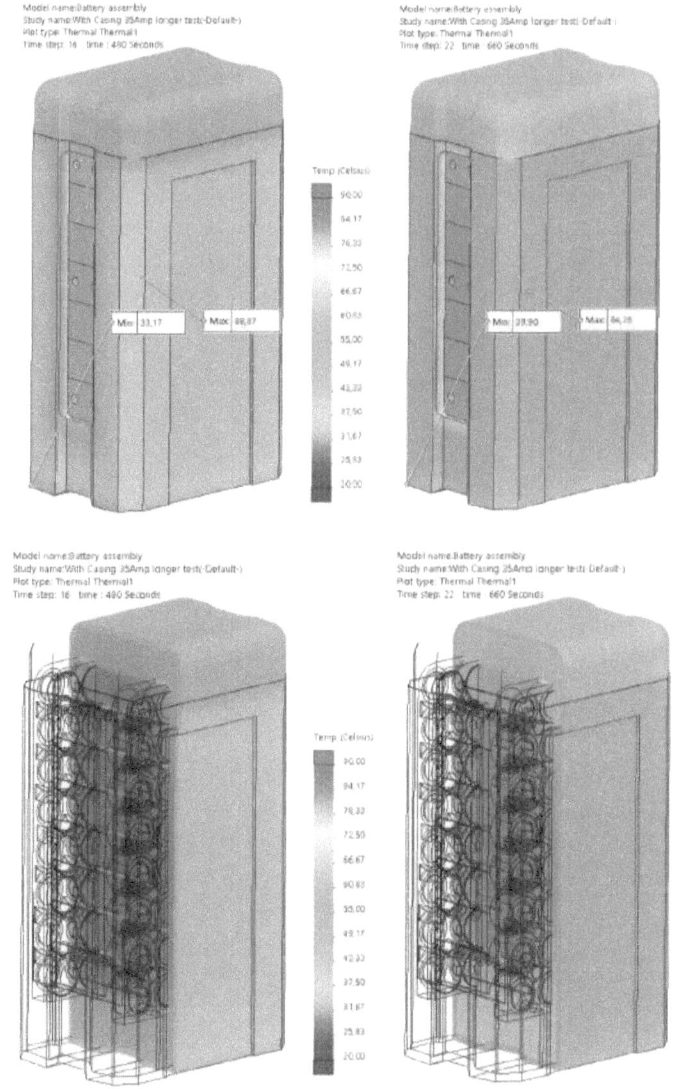

Figure 36: Left: Simulation result after 480 s, Right: Simulation result after 660 s.

4.4 Comparison of Simulation v. Testing

The simulation results are now compared to the real-life discharge tests at different positions and different currents. *Figure 37(a-c),* compare the simulation to the 35 Amps tests without casing, while *Figure 37 d-g),* compare the simulation at 45 Amps without casing. Between these two simulations, the only change was the value for heat power, just like in real-life tests, where the only element changed was the current load.

Figure 37(a-g), show the position measured on the aluminum heat sinks at 35 & 45 Amps without casing, where the red dotted line represents the results from the simulation. The position from the aluminum heat sinks is an excellent representation of the rate the battery is heating up in real-life. What becomes more evident in the 45 Amps tests, is the slow ramp-up of the temperature compared to the simulation value. This slow ramp-up is most like due to the battery in real-life changing heat power value as the load is applied where the simulation is using one value for heat power throughout the entire heating phase.

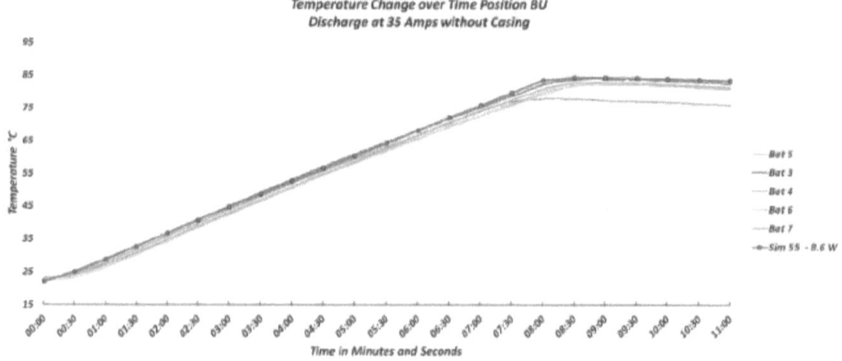

Figure 37: Comparison of simulation result vs. test result at BU – 35 Amps

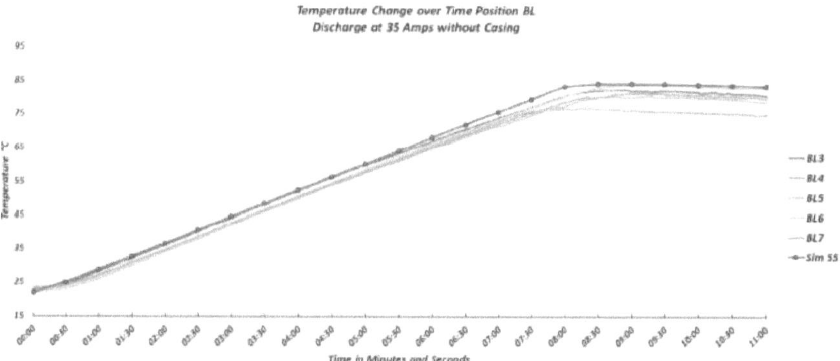

Figure 37 a: Comparison of simulation result vs. test result at BL – 35 Amps

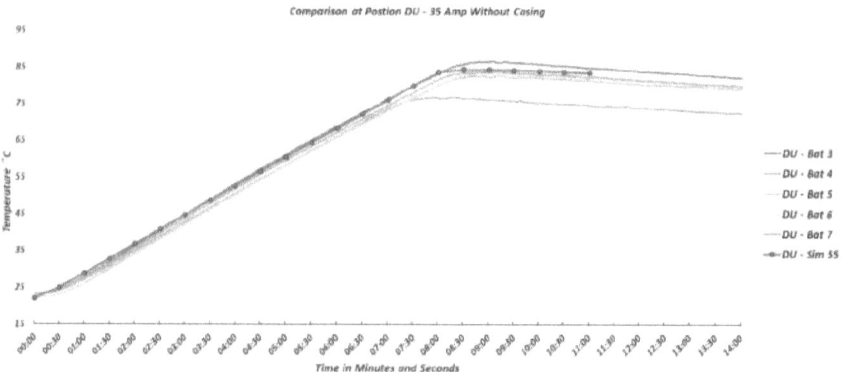

Figure 37 b: Comparison of simulation result vs. test result at DU – 35 Amps

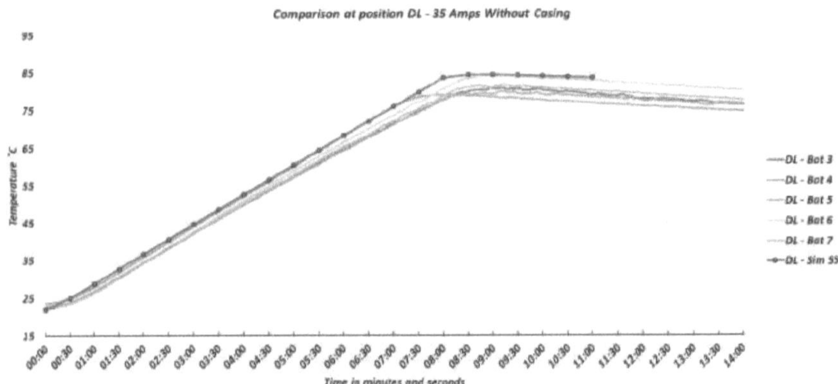

Figure 37 c: Comparison of simulation result vs. test result at DL – 35 Amps

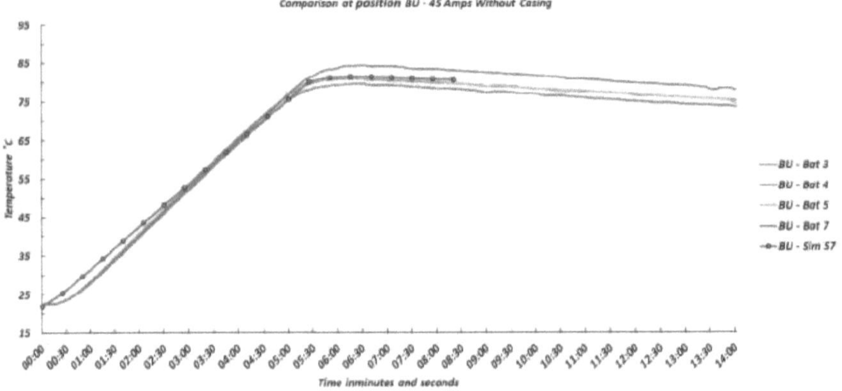

Figure 37 d: Comparison of simulation result vs. test result at BU – 45 Amps

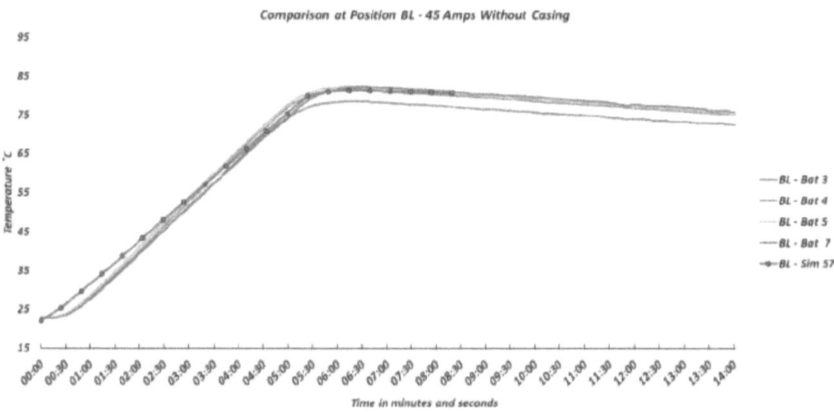

Figure 37 e: Comparison of simulation result vs. test result at BL – 45 Amps

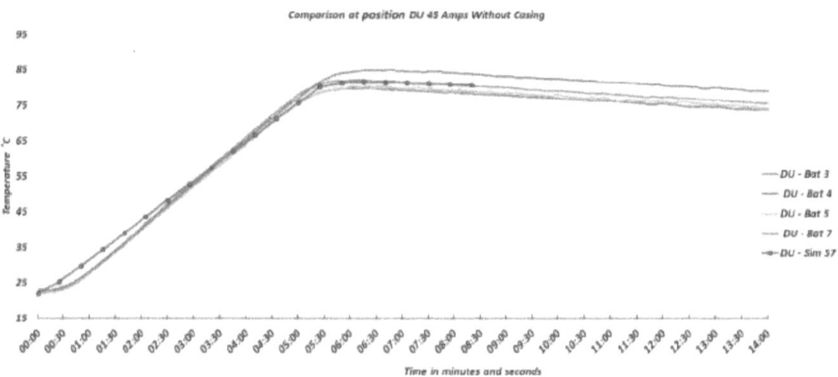

Figure 37 f: Comparison of simulation result vs. test result at DU – 45 Amps

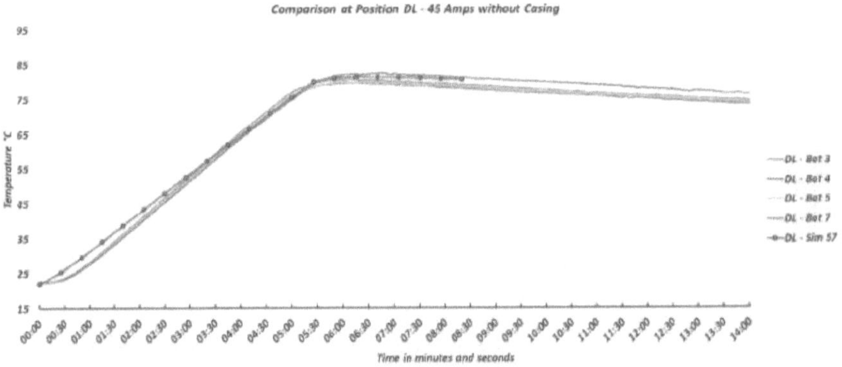

Figure 37 g: Comparison of simulation result vs. test result at DL – 45 Amps.

Figure 38 shows the simulation results for the internal sensor at 35 Amps without casing. These results match quite well, which is essential since this location represents the sensor that shuts-off the battery once it reaches the shutdown temperature. Matching the temperature at this location can then provide insight into how long the battery pack can run before the software shuts it off.

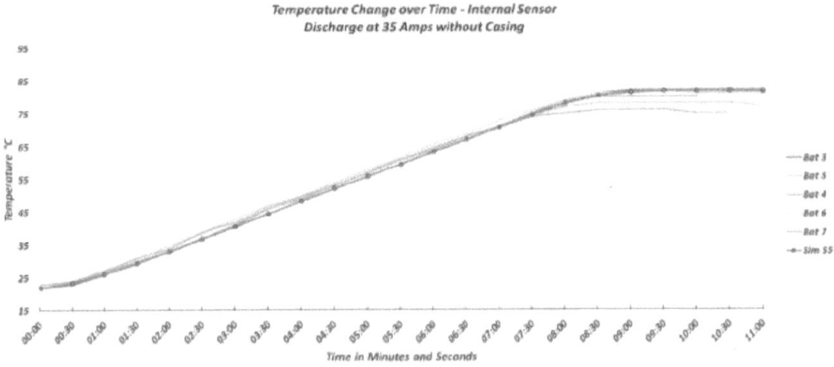

Figure 38: Comparison of simulation result vs. test result of internal sensor – 35 Amps without casing.

Figure 39 shows the results from the silicone sleeve, the material that protects the battery from being penetrated, and battery swelling. The simulation result is on a steeper slope and ending at a higher temperature. This slope is most likely due to the incorrect material properties. The BOM from manufacturing did confirm the material to be a type of silicone and that the thermal conductivity is at least 2 *W/m.C.* The BOM suggests that the thermal conductivity used in the simulation may be too low. An optimization can be performed to improve the results.

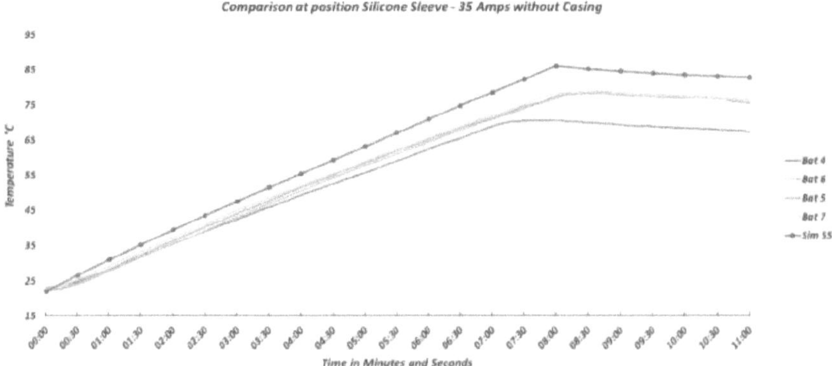

Figure 39: Comparison of simulation result vs. test result at silicone – 35 Amps.

Figure 40 (a-c), compare the positions for the ends of the battery cells. As seen before, the thermal behavior of the lithium cells is unique, which would require more research to recreate the same behavior in the simulation.

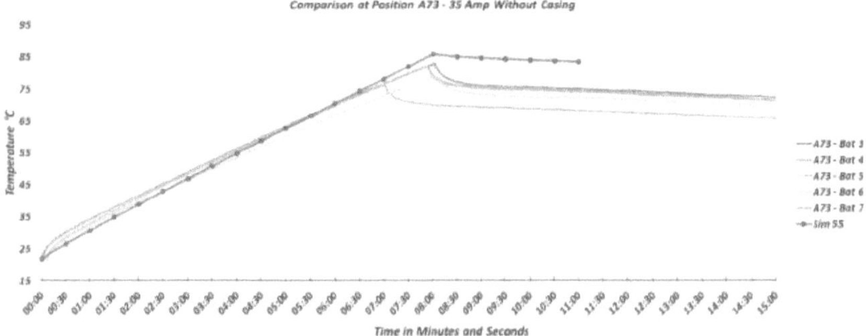

Figure 40: Comparison of simulation vs. test at position A73 – 35 Amps.

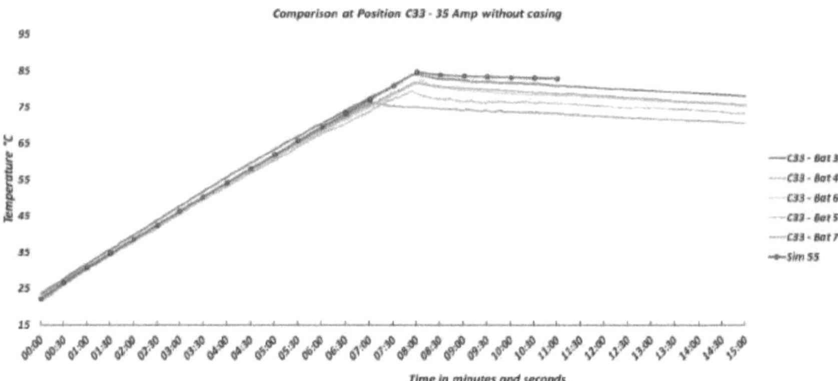

Figure 40 a: Comparison of simulation vs test at position C33 – 35 Amps.

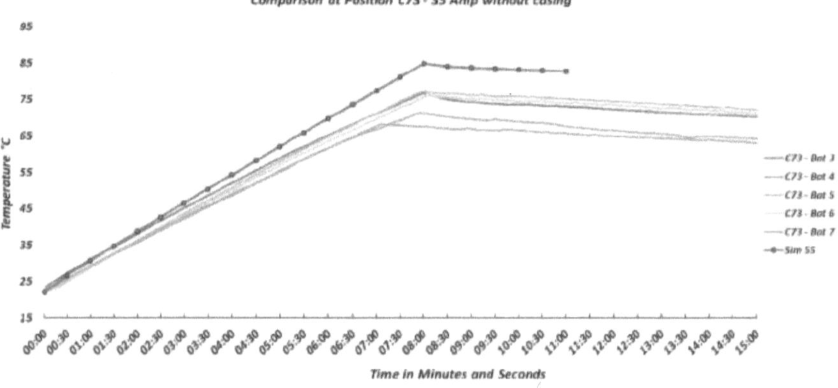

Figure 40 b: Comparison of simulation vs test at position C73 – 35 Amps.

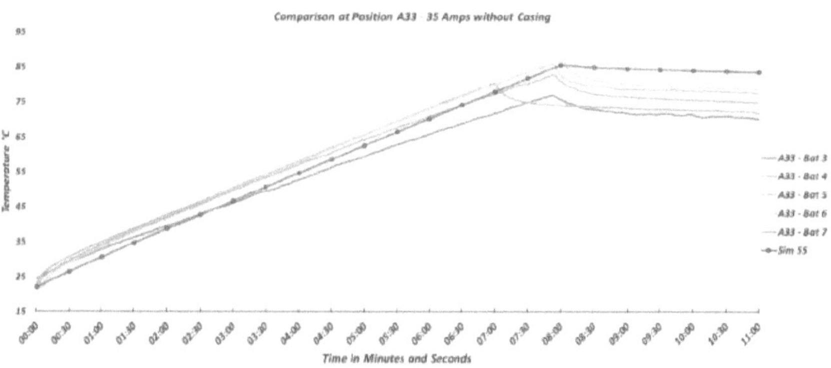

Figure 40 c: Comparison of simulation vs test at position A33 – 35 Amps without casing.

Figure 40 d-*h* are the results from the 35 Amps test with the casing on in the simulation. The results comparisons for positions on the outer casing (BU & BL) and the aluminum (AL & CL) match with the experimental results quite well. These results suggest that the method used to modeled air as a component worked quite well. However, *Figure 40 h* for the top cover of the battery pack was way off, with the simulation results being much higher than what was measured. These results could be due to incorrect conduction contact value, incorrect material properties for the top cover, or incorrect geometry for the top cover.

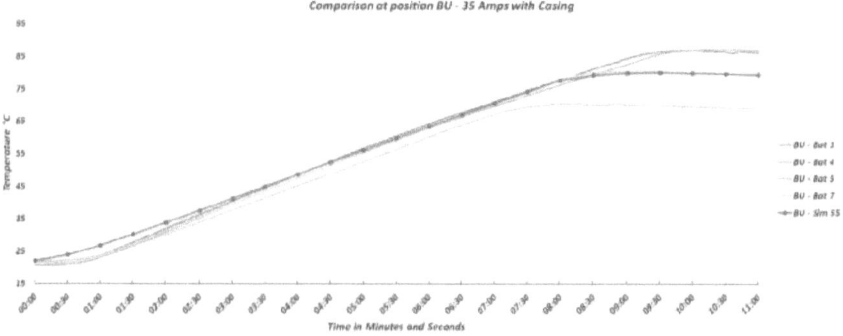

Figure 40 d: Comparison of simulation vs test at BU – 35 Amps with casing.

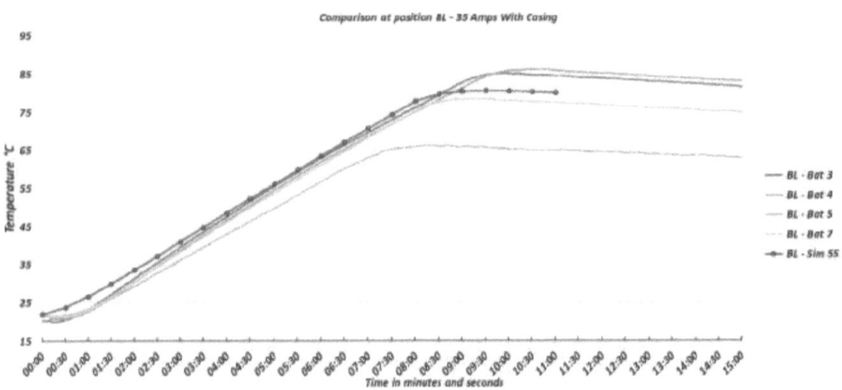

Figure 40 e: Comparison of simulation vs test at BL – 35 Amps with casing

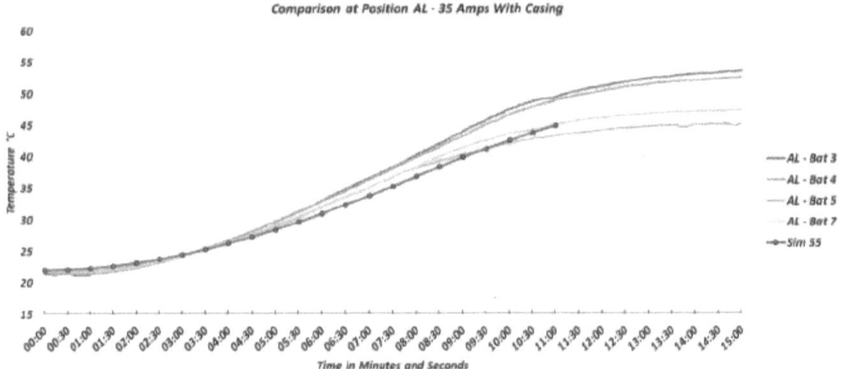

Figure 40 f: Comparison of simulation vs test at AL – 35 Amps with casing

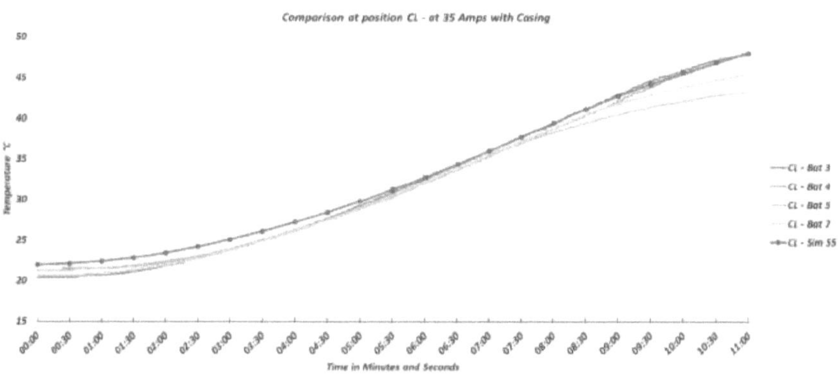

Figure 40 g: Comparison of simulation vs test at CL – 35 Amps with casing

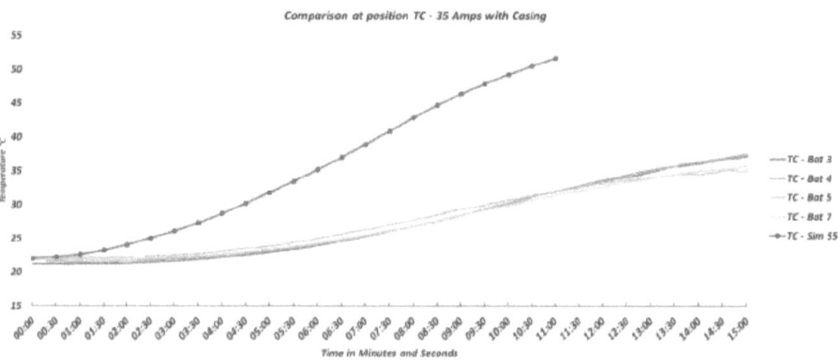

Figure 40 h: Comparison of simulation vs test at TC – 35 Amps with casing

4.5 Redesign Suggestions

This paper investigated the heat dissipation of the lithium-ion battery pack under different discharge rates and offered a simulation model for Greenworks to use in their next generation of battery packs. Throughout this investigation, a few factors, such as cell arrangement, heatsink design, and the battery pack encloser, were found to hinder the battery pack's cooling ability. These factors should be considered in the next generation of the battery pack to improve the performance.

4.5.1 **Cell Arrangement:**

When opening the Cramer battery pack, the arrangement of the cells is the first thing noticed. The battery cells are tightly packed together in an 8X5 rectangular arrangement. This configuration reduces the battery pack volume but also limited the amount of exposed surface area of each cell. The low amount of surface area leaves little to no air circulation pass between the cells and cools the batteries in the middle. The aluminum heat sink does help in distributing the heat throughout the system, but spacing the batteries out would help to dissipate the heat faster. While aluminum helps to keep the battery within the operating temperature for a longer amount of time, it takes far too long to remove the heat from the aluminum. What may help to reduce the time for cooling is to rearranging battery cells to provide better air circulation inside the battery and more surface area to dissipate the heat. Possible rearrangements are in the *Figure 41*. As a comparison for size, the cell arrangement on the far right is the arrangement of the cells inside the Cramer battery pack.

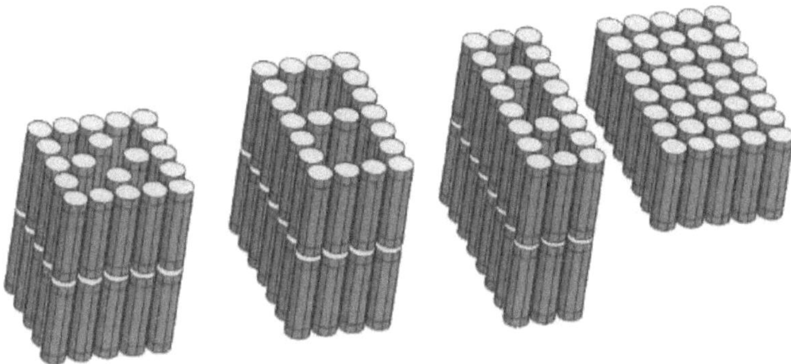

Figure 41: Different cells configurations.

4.5.2 **Battery Enclosure**

The *IP* code or Ingress Protection Code describes the level protection the internal components have to the outside environment. The Cramer battery pack has an ingress level of *IPX4*, which means the cells inside the product are protected against splashing water. This protection is an excellent feature to describe how the battery can be used in different weather conditions, but this feature is hurting the performance of the products. The *IPX4* prevents most water from entering and reduces the air circulation, which causes the rise of temperature inside the battery, thereby affects battery performance. Lithium-ion batteries generate a significant amount of heat during high current discharge; therefore, a larger heat sink or an active cooling system should be considered. The Cramer battery is using a passive cooling system, where only the aluminum heat sinks dissipate heat. The current heatsinks cannot dissipate the amount of heat that cells generate. The mass and exposed surface area of the aluminum needs to increase to help dissipate heat. The other solution is using an active cooling system to allow moving fluid, either gas or liquid, to adsorbed and dissipate the heat from the cells. Both solutions may cause the battery to fail the *IPX4* requirement and change the connection interface. Choosing to make the battery meet the *IPX4* requirement or improve the cooling system is up to Greenworks to decide.

4.5.3 **Phase Changing Materials (PCM):**

The use of Phase Changing Materials (PCM) materials in thermal management of lithium-ion batteries has increased recently, due to the material's high latent of fusion heat. This material requires a high amount of energy during the melting and solidifying process. PCM absorbs heat from the surrounding environment until it reaches the melting point. The absorbing process continues until the whole PCM material is melted. This energy stores inside the PCM as latent heat and will releases to the surrounding environment as the temperature decreases, and the PCM starts the solidifying process, as shown in *Figure 42* [34]. Considering PCM offers many different temperature ranges and has high-density heat energy storage, this material makes as a good alternative for heat sinks in this type of batteries.

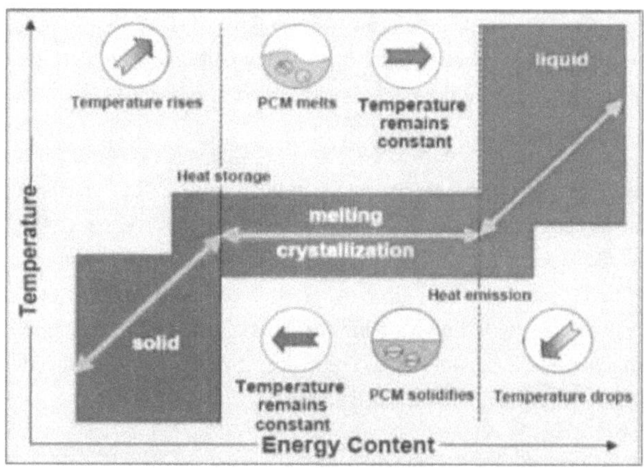

Figure 42: Principle of phase change material [35].

5 Discussion and conclusions

This section summaries and describes the findings and answer the questions stated in the purpose. Future projects that can continue the work of this book or aid in other areas are described as well.

5.1 Discussion of method

The phases of the book were to research how to conduct the discharge test, collecting thermal data from tests, setting up a simulation for the battery pack, and finally calibrating the simulation to match what was occurring in real-life. The approach used was quite smooth and did not require much backtracking but took longer than expected.

The methodology behind the approach was to help create a tool to design a new battery pack. There are many solutions to solve heating issues that could have been prototyped and tested straight away, but this would not answer if the solution is the best for the Greenworks. The best solution for a company is usually the one that makes the most profit, which may be the cheapest, lightest or, the most powerful solution. By simply answering the overheating question, it would not answer if the proposed solution is the best for the company. Instead of proposing a solution to the overheating issue, this book proposes a tool for Greenworks to use. This simulation tool gives Greenworks' the opportunity to test all their ideas before manufacturing or prototyping anything, saving time and reducing cost.

Once this simulation approach was set, the next issue was to ensure the simulation results would be realistic. This was solved by testing the battery packs at the current output to represent the performance Greenworks would like to have for the battery pack. Before the testing, research for how the type of test should be conducted was done and found that there were no set standards for this type of test. The Greenworks branch had performed this same test before and shared the procedure. The design of the test was similar to those found in the research, which assured this method for testing would be acceptable. The testing procedure can be found in *section 3.2.1*, and the equipment used is listed in *appendix 7.1*.

The results collected from the experiment were entirely consistent between all the batteries when they were compared to each other. The rate of temperature rise was quite linear as well, which suggested that the heat power coming from the batteries was steady and constant. This also meant it would be much easier to recreate in the simulations.

The manufacturing of the batteries was significant to the experimental data since any alteration between them would make the data useless. During the experiments, it was discovered that batteries 1 and 2 had two additional heat sinks the other five battery packs did not have. This inconsistency meant that the data from these two batteries could not be used in the comparisons or the simulations.

During the testing and data collection, the simulation of the battery was being set up and tested. The simulation type was quite different, so smaller scale simulation was used at first to understand how to use the simulation. The model from Greenworks was also challenging to turn into a usable CAD model in Solidworks. Eventually, both CAD model and thermal simulation for Solidworks were under control and were able to be used for a larger scale. Understanding how to appropriately use Solidworks simulation and applying the parameters took the most amount of time. One simulation method that was considered but not used was symmetrical study. Since the battery pack is symmetrical about the center front face, the thermal study could have been simulation on just one side and copied over. This method would have saved time in computing the results and set up of the model. This method was not used due to fear that the result would not be simulated correctly.

The simulation phase also included researching the three forms of heat transfer and how the lithium-ion batteries are generating heat power. It was discovered during this research how to measure the heat power being released from the lithium-ion cell based on the chemical reaction. The measurement required the cell temperature, battery state of charge (SOC), and the battery open-circuit voltage as the battery was discharged. This measurement method also only focused on testing one battery cell at a time, which was not possible for the given battery pack. This research for the battery cell heating should have been done prior to testing so that the data need to measure the chemical reaction could have been collected. This also led to spending a lot of time analyzing the simulation parameters to determine why the simulation results could not match what was collected from the experiments.

The comparison of the data was the most exciting phase of the book, from the start comparing the experiment results at the same location and then adding the simulation results. Observing how well the experiment data match each other was a good sign; the testing method was sound. Overall, the method used to develop the simulation parameters went quite smoothly and provided excellent results.

5.2 Discussion of findings

In this section, various aspects of this book are discussed, with regard to the critical decisions that made it possible to carry out this project, it also addresses the limitations that were encountered during the project.

Before the research and testing began, Greenworks essentially had nothing to show for testing and simulation for this battery pack; this meant that all work was from scratch. The bulk of the work for this book was with the testing and determining the stimulation parameters.

Researching the testing standards prove to be quite tricky since most companies that provided testing standards required payment to use their tests. Fortunately, some of the testing companies did respond to our emails about temperature testing lithium-ion batteries, but none had a testing standard for the test we described. The few tests we were able to access did have some testing details but did test what we wanted to. Since none of the companies we contacted knew of a test that measured temperature during the discharge test, we had to use the testing procedures provided by Greenworks.

The execution of the tests was critical because once we started the test, we could not stop it, and we could not repeat a test on a battery in the same day. Since we are draining the energy from the battery pack, we cannot stop the test and restart it without recharging the battery. The pressure to perform the test correctly was double because we also measured the temperature of the battery pack if we either charged or discharged the battery the temperature would increase and have an incorrect starting temperature for the pack.

During testing, we discovered that mounting the sensors on the battery pack proved to be a little difficult. When the battery heated above 50 degrees C, the mounting tape would peel, and the sensors come loose. We could not permanently secure the sensors because we had a limited number, and they were relatively expensive. While researching for tests, we found that thermal runaway test procedure call to glue thermocouples on to the testing surface. We learned that thermocouples are much cheap but less accurate. This led us to believe thermal runaway tests use the cheaper sensors that are glued in place with thermal paste because the sensors are destroyed during the testing. We copied the ideas to use the thermal paste to help hold the sensors and get finer temperature readings. We also changed the tape used for sensors for something that could withstand the heat of the battery pack. With these changes, the sensor had no more issues.

Another issue with managing the testing was synchronizing the data between the 3 individual temperature data collectors, the discharge machine, the thermal camera, and the voltmeter. All these testing tools operated on independent systems, and we had to make sure the data collected from them were compared at the same starting time. Synchronizing the data was a requirement to be able to compare the temperature data between the different battery packs. We synchronized the tools using the local time of the area. For 3 Unilogs (temperature collectors), we could set the internal clock to the nearest second, and this time was attached to the output data. For the discharge machine, we used a household clock synchronized with the Unilogs, and coordinated when to start the discharge machine. For the thermal camera and voltmeter, the house clock was used again. When we took photos with the thermal camera, we made sure to include the clock in the photo. The data from the voltmeter was video recorded with the clock in view. These methods were a bit crude, but since the test duration was quite long, we did not notice a significant issue with the data when compared to each other.

The parameters for the thermal simulation most dealt with researching, calculating, and testing different parameter combinations. The parameters for the conduction were straight forward to determine just required research for the material properties and calculating the thermal resistance. The convection value was far more challenging and required some assistance from David, who suggested to use the HVAC study to determine the convection coefficients.

This book also confirmed the difficulty in accurately determining the heat power being generated from the lithium-ion cells. Much time was spend researching the methods for calculating the heat power only to discover we could not calculate it with the data we had. The heat capacity method for calculating the heat power was an excellent method to support the test data and support the findings that a large percentage of the heat power from the battery pack was due to the chemical reaction occurring inside the battery. We think the heat capacity method is quite accurate in determining heat power value.

The limitation of the simulations is that it can only be used to simulation battery packs using the same lithium-ion batteries. Knowing that a large percentage of the heat power comes from the chemical reaction occurring inside the cell, batteries of different chemical configurations may produce a different amount of heat power. This limitation also applies to the discharge current going through the battery cells. These simulations results can only be applied to simulate a constant current discharge at 35 or 45 Amps. The heat power values used are based on the experiment results collected, to be able to simulate the battery pack being discharged at different loads, then experiments at that load must be conducted. This issue may be overcome by using the heat capacity method and overestimating the rate the temperature is rising, thereby producing a larger heat power value. Keeping in mind, these simulation parameters are used to help develop the next battery. The simulation should provide a small safety factor to assure that the real-life battery does not overheat. So, by having a larger heat power value, you will be adding a slight safe factor for the temperature of the battery pack.

The value of the results from this book is the simulation model and testing method to collect the data. The simulation can be used for the future development of the battery pack, and the testing model could be used to test other batteries.

5.3 Conclusions

The primary purpose of the book was to help Greenworks develop their next battery pack. This was done by creating a simulation environment for the new battery design to be tested within. A simulation environment in place of constructing prototypes can save manufacturing costs and time for developing and testing ideas. In the process of developing the simulation environment, the following questions that were laid out, in the beginning, were answered.

- Why is the battery pack overheating?

The work in this book found that the battery pack is overheating because of the high load and insufficient thermal management. The high current passing through the cell generates heat much faster than the passive heat sink can it dissipated. The heat itself is generated through ohmic heating and the exothermic chemical reaction occurring inside the cells. The ohmic heating is based on the electrical resistance and the applied current, even though the lithium-ion cells have very little internal resistance, the electrical current is so high that a significant amount of heat is generated.

- What is the procedure to conduct valid thermal testing on this battery pack?

There is not an officially recognized testing standard to thermal test lithium batteries without the intent to cause thermal runaway. The testing procedure used was based on standards to test batteries for thermal runaway, cell penetration, and Greenworks manufacturing test. Based on the quality of the results collected and comparing them to the sensors manufacturers used inside of the battery pack, the test design was a valid method.

- What are the simulation parameters to recreate the results collected from the battery discharge temperature testing?

The simulation parameters for recreating the experiment results are listed out in *section 3.3*. Based on the comparison between the simulation and the experiment, the parameters used should be accurate enough to aid in the development of future design, assuming the parameters are applied correctly.

- Can the simulation's parameters be used for other tests or products?

The simulation parameters are limited to the experiments they have been designed for, but some aspects can still be used. They are limited because the heat power values are based on experiment results. The methods for calculating the values for convection and conduction are still valid to use for other tests or products.

- What are some suggestions for the redesign to meet the thermal requirements?

The result of this study recommends that increased fluid flow through the aluminum heat sink and battery cell would significantly help in the cooling of the battery pack.

5.4 Future Work

The finding in this book could be followed up in a few different directions:

- The calculation for the heat power being generated is still not predictable. Further work can be done to develop a formula to calculate the heat power based on the current being applied to the battery pack.

- This simulation set up is ready to be used to develop the next generation of the battery pack, this could be used to test different design and material for the battery pack

- The simulation parameters need to be improved to model the temperature curves in a couple of areas. Work can be done to identify what is the issues are and fix them to improve the simulation accuracy. The areas that need work are positive and negative end of the battery cells, the top cover, and the silicone sleeve.

- A flow simulation can be created to test how the air is behaving inside the battery pack. The flow simulation could be used to show how increasing the airflow inside the battery pack can affect cooling performance.

6 References

[1]   A. Matrins, "Most Consumers Want Sustainable Products and Packaging," Business News Daily, 4 June 2019. [Online]. Available: https://www.businessnewsdaily.com/15087-consumers-want-sustainable-products.html. [Använd 22 07 2020].

[2]   T. Whelan och R. Kronthal-Sacco, "Research: Actually, Consumers Do Buy Sustainable Products," Havard Business Review, 19 June 2019. [Online]. Available: https://hbr.org/2019/06/research-actually-consumers-do-buy-sustainable-products. [Använd 22 07 2020].

[3]   Cramer, "82V220-82V430 CRM 82V Battery Manual," [Online]. Available: https://cramertools.com/sites/cramer/files/2019-03/82V220-82V430%20CRM%2082V%20Battery%20Manual_15.pdf. [Använd 22 07 2020].

[4]   Sony, "Data Sheet for Sony US18650VTC6," 30 06 2015. [Online]. Available: https://www.18650batterystore.com/v/files/sony_vtc6_data_sheet.pdf. [Använd 14 01 2020].

[5]   M.-d. Li och F. Wang, "Thermal performance analysis of the lithium-ion batteries," 22010.

[6]   N. H. F. Ismail, S. F. Toha, N. A. M. Azubir, N. H. M. Ishak, M. K. Hassan och B. S. K. Ibrahim, "Simplified Heat Generation Model for Lithium-ion battery used in Electric Vehicle.," i *International Conference on Mechatronics (ICOM'13)*, Kuala Lumpur, 2014.

[7]   D. Linden och T. B. Reddy, Handbook of batteries, New York: McGraw Hill, 2001.

[8]   E. Winker, "Lithium-ion Battery Aging: Battery Lifetime Testing and Physics-based Modeling for Electric Vehicle Application," 2017.

[9]   Sony, Lithium-ion rechargeable Batteries, Technical Handbook, Osaki west Technology Center.

[10]   Battery university, "https://batteryuniversity.com/," 10 07 2019. [Online]. Available: https://batteryuniversity.com/learn/article/how_to_prolong_lithium_based_batt eries. [Använd 10 04 2020].

[11]   L. Lam, "A Practical Circuit based Model for State of Health Estimation of Li-ion Battery Cells in Electric Vehicles," 2011.

[12]   S. S. Choi och H. S. Lim, "Factors that affect cycle-life and possible degradition mechanisms of a Li-ion cell based on LiCoO2," *Journal of power sources,* 2002.

[13]   M. Murnane och A. Ghazel , "A Closer Look at State of Charge (SOC) and State of Health (SOH) Estimation Techniques for Batteries," 2017.

[14]   S. Du, Y. Lai, L. Ai, Y. Cheng , Y. Tang och M. Jia, "An Investigation of irreversible heat generation in lithium-ion batteries based on a thermo-electrochemical coupling method," *Applied Thermal Engineering,* 2017.

[15]   N. Sato, "Thermal behavior analysis of lithium-ion batteries for electric and hybrid vehicles," *Journal of power sources,* vol. 99, pp. 70-77, 2001.

[16]   X. Zhang, "Thermal analysis of a cylindrical lithium-ion battery," *Electrochimica Acta,* 2010.

[17]   R. Benger, H. Wenzl, H.-P. Beck, M. Jiang, D. Ohms och G. Schaedlich, "Electrochemical and thermal modeling of lithium-ion cells for use in HEV or EV application," *World Electric Vehicle Journal*, 2009.

[18]   D. H. Jeon och S. M. Baek, "Thermal modeling of cylindrical lithium ion battery during discharge cycle," *Energy Conversion and Management*, 2011.

[19]   F. P. Incropera och D. P. Dewitt, Introduction of heat transfer, Jefferson City: John Wiley & sons,INC.

[20]   J. H. Lienhard IV och J. H. Lienhard V, A Heat Transfer textbook, Cambridge: Phlogiston Press, 2019.

[21]   Y. A. Cengel, Heat Transfer: A practical Approach, Mcgraw-Hill (Tx), 2002.

[22]   S. A. Klein och F. L. Alvarado, *Properties of air at 1 atm pressure*, 2012.

[23]   J. Bird, Electrical Circuit Theory and Technology, Oxford: Newnes, 2001.

[24]   D. H. Doughty, *The Landscape of Thermal Runaway Propagation*, 2018.

[25]   S. R. Singiresu, The Finite Element Method in engineering, Burlington: Elsevier Butterworth, 2005.

[26]   Underwriters Laboratories, "ANSI/CAN/UL Standard for Test Method for Evaluating Thermal Runaway Fire Propagation in Battery Energy Storage Systems," 12 11 2019. [Online]. Available: https://www.shopulstandards.com/ProductDetail.aspx?UniqueKey=36503. [Använd 22 01 2020].

[27]   United Nations, "Uniform provisions concerning the approval of vehicles with regard to specific requirements for the electric power train," United Nations, 2011.

[28]   NASA, "Crewed Space Vehicle Battery Safety Requirements," Lyndon B. Johansson Space Center, Houston, 2017.

[29]   C. J. Orendorff, J. Lamb och L. A. M. Steele, "Recommended Practices for Abuse Testing Rechargeable Energy Storage Systems (RESSs)," Sandia National Laboratories, Albuqerque, 2017.

[30]   L. Peeters, I. Beausoleil-Morrisom, B. Griffith och A. Novoselac, "Internal convectioncoefficients for building simulation," *Proceedings of building simulation,* 2011.

[31]   X. Wang, X. Wang, G. Xing, J. Chen, C.-X. Lin och Y. Chen, "Intelligent Sensor Placement for Hot ServerDetection in Data Centers," 2012.

[32]   N. S. Spinner, K. M. Hinnant, R. Mazurick, A. Brandon och S. L. R. Pehrsson, "Novel 18650 lithium-ion battery surrogate celldesign with anisotropic thermophysical propertiesfor studying failure events," *Power Sources,* 2016.

[33]   A. Lebkowski, "Temperature, Overcharge and Short-Circuit Studies of Batteries used in Electric Vehicles," pp. 67-73, 2017.

[34]   PureTemp, "https://www.puretemp.com," PureTemp, [Online]. Available: https://www.puretemp.com/stories/understanding-pcms. [Använd 15 May 2020].

[35]   K. Almadhori och S. Khan, "A review - An Optimazation of Macro-encapsulated Paraffin used in Solar Latent Heat Storage Unit," *International Journal of Engineering and Technical Research (IJERT),* vol. 5, nr 01, p. 9, 2016.

7 Appendices

7.1 Testing Equipment

The testing equipment includes the major tools and devices use to conduct the experiments and the simulations. All the testing equipment was provided by Greenworks except for the CAD and simulation software.

7.1.1 Battery Testing Load

To have valid and consistent tests the same loading conditions has been applied to each battery pack while recording the required data. To meet the load testing requirement there were three available options:

1. Use the battery pack normally inside of a Greenworks chainsaw and recreate the overheating issue as presented by Greenworks.
2. Create a test rig from a power tool, which can be operated inside a controlled environment.
3. Use a discharge machine, which can drain the power from the battery pack at a constant current, voltage, or custom load.

It most closely met the experiment requirements would provide the most consistent results. This method is also shared with most of the international standards and the tests designed by the Greenworks lab in China. The discharge machine provided by Greenworks is the *"Elekro-Automatik: ELR 9000 HP Electronic DC Load with Energy Recovery"*, seen in *Figure 43*. This machine can discharge the battery packs at constant or non-constant current profile and recorded the batteries voltage drop over time. Having full control over load profile it will ensure that each battery pack will undergo the exact same testing process. The other benefit of this machine is that it will be much easier to control the battery testing environment such as airflow, temperature, and equipment. By having close control over the loading and environmental affects producing good data and simulations will be much easier and more reliable.

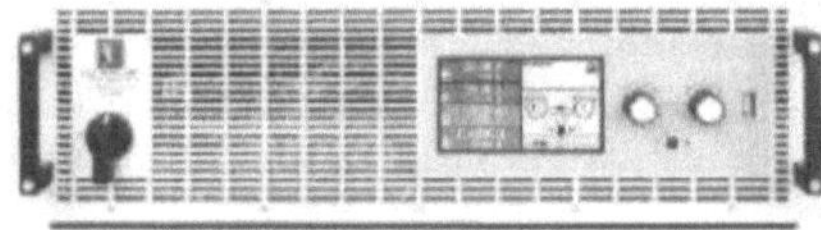

Figure 43: ELR 9000 HP Electronic DC Load

81

7.1.2 **Data Recorded**

In order to compare the results of the tests versus simulation, data from the experimental tests needed to be recorded. The data was collected with the *Unilog 2* data collector. Unilog is a small data collector device that has many inputs to measuring current, voltage, power, capacity, and temperatures *Figure 44*. The data can be collected into an SD card and be viewed on the small display (Unidisplay), which allows the experiment to be constantly monitored by reading the data and abort the experiments if the temperatures get too high.

Figure 44: Unilog

7.1.3 **Thermal Camera**

To get an overview of the temperature conditions and to visualize the temperature of the battery pack during the discharge test, a thermal camera was used. The thermal camera, which also calls infrared camera, measures infrared radiation from the objects and convert it into temperatures. For conducted experiments the *FLIR R2-Cx* thermal camera was used shown in *Figure 45*. This thermal camera has range of *-10* to *150 C°* and automatically find the hotspots of the object.

Figure 45: Thermal camera FLIR model: R2-Cx

7.1.4 **Calibration of equipment**

In order to ensure the accuracy and repeatability of sensors, all *PT-1000* sensors needed to be calibrated. Calibration is a process that constitutes for those machines that do not meet the specified requirements, where an adjustment is made on a sensor to make them work as accurately or as error-free as possible, to achive that *Table 11* have been used to calibrate all *PT-1000* sensors. Proper calibration enables an exact and precise measurements which in turn gives better control of the test.

Having to calibrate sensors does not always mean that the sensor does not meet the requirements, but may be that the process operating range is changed or the sensors have been subjected to extreme temperatures, do not have the same zero reference or spilled liquids that can affect the accuracy of the sensors.

°C	0,0	-1,0	-2,0	-3,0	-4,0	-5,0	-6,0	-7,0	-8,0	-9,0
-50,0	803,1									
-40,0	842,9	838,8	834,8	830,8	826,9	822,9	818,9	815,0	811,0	807,0
-30,0	822,2	878,3	874,3	870,4	866,4	862,5	858,5	854,6	850,6	846,7
-20,0	921,6	917,7	913,7	909,8	905,9	901,9	898,0	894,0	890,1	886,2
-10,0	960,9	956,9	953,0	949,1	945,2	941,2	937,3	933,4	929,5	925,5
0,0	1000,0	996,1	992,2	988,3	984,4	980,4	976,5	972,6	968,7	964,8

°C	0,0	1,0	2,0	3,0	4,0	5,0	6,0	7,0	8,0	9,0
0,0	1000,0	1003,9	1007,8	1011,7	1015,6	1019,5	1023,4	1027,3	1031,2	1035,1
10,0	1039,0	1042,9	1046,8	1050,7	1054,6	1058,5	1062,4	1066,3	1070,2	1074,0
20,0	1077,9	1081,8	1085,7	1089,6	1093,5	1097,3	1101,2	1105,1	1109,2	1112,8
30,0	1116,7	1120,6	1124,5	1128,3	1132,2	1136,1	1139,9	1143,8	1147,7	1151,5
40,0	1155,4	1159,3	1163,1	1167,0	1170,8	1174,7	1178,5	1182,4	1186,2	1190,1
50,0	1194,0	1197,8	1201,6	1205,5	1209,3	1213,2	1217,0	1220,9	1224,7	1228,6
60,0	1232,4	1236,2	1240,1	1243,9	1247,7	1251,6	1255,4	1259,2	1263,1	1266,9
70,0	1270,7	1274,5	1278,4	1282,2	1286,0	1289,8	1293,7	1297,5	1301,3	1305,1
80,0	1308,9	1312,7	1316,6	1320,4	1324,2	1328,0	1331,8	1335,6	1339,4	1343,2
90,0	1347,0	1350,8	1354,6	1358,4	1362,2	1366,0	1369,8	1373,6	1377,4	1381,2
100,0	1385,0	1388,8	1392,6	1396,4	1400,2	1403,9	1407,7	1411,5	1415,3	1419,1
110,0	1422,9	1426,6	1430,4	1434,2	1438,0	1441,7	1445,5	1449,3	1453,1	1456,8

Table 11: the resistance related to the temperature. applicable platinum elements PT1000

7.2 Table: Natural convection over surfaces

Empirical correlations for the average Nusselt number for natural convection over surfaces

Geometry	Characteristic length L_c	Range of Ra	Nu
Vertical plate 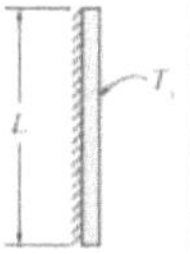	L	10^4–10^9 10^9–10^{13} Entire range	$Nu = 0.59Ra_L^{1/4}$ (9-19) $Nu = 0.1Ra_L^{1/3}$ (9-20) $Nu = \left\{0.825 + \frac{0.387Ra_L^{1/6}}{[1 + (0.492/Pr)^{9/16}]^{8/27}}\right\}^2$ (9-21) (complex but more accurate)
Inclined plate 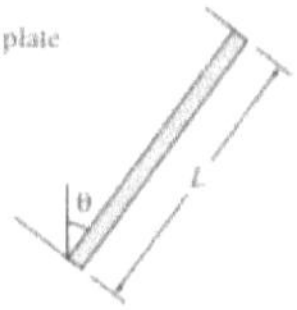	L		Use vertical plate equations for the upper surface of a cold plate and the lower surface of a hot plate Replace g by $g\cos\theta$ for $Ra < 10^9$
Horizontal plate (Surface area A and perimeter p) (a) Upper surface of a hot plate (or lower surface of a cold plate)  (b) Lower surface of a hot plate (or upper surface of a cold plate) 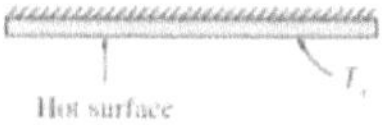	A_s/p	10^4–10^7 10^7–10^{11} 10^5–10^{11}	$Nu = 0.54Ra_L^{1/4}$ (9-22) $Nu = 0.15Ra_L^{1/3}$ (9-23) $Nu = 0.27Ra_L^{1/4}$ (9-24)
Vertical cylinder 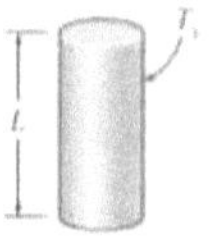	L		A vertical cylinder can be treated as a vertical plate when $D \geq \frac{35L}{Gr_L^{1/4}}$
Horizontal cylinder 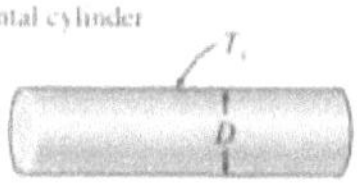	D	$Ra_D \leq 10^{12}$	$Nu = \left\{0.6 + \frac{0.387Ra_D^{1/6}}{[1 + (0.559/Pr)^{9/16}]^{8/27}}\right\}^2$ (9-25)
Sphere 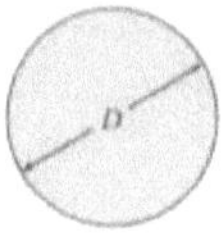	D	$Ra_D \leq 10^{11}$ $(Pr \geq 0.7)$	$Nu = 2 + \frac{0.589Ra_D^{1/4}}{[1 + (0.469/Pr)^{9/16}]^{4/9}}$ (9-26)

Table 12: Natural convection over surfaces

7.3 Table: Emissivity of common material

TABLE A.8 Total, Normal (*n*) or Hemispherical (*h*) Emissivity of Selected Surfaces

Metallic Solids and Their Oxides[a]

Description/Composition		Emissivity, ε_n or ε_h, at Various Temperatures (K)										
		100	200	300	400	600	800	1000	1200	1500	2000	2500
Aluminum												
Highly polished, film	(*h*)	0.02	0.03	0.04	0.05	0.06						
Foil, bright	(*h*)	0.06	0.06	0.07								
Anodized	(*h*)			0.82	0.76							
Chromium												
Polished or plated	(*n*)	0.05	0.07	0.10	0.12	0.14						
Copper												
Highly polished	(*h*)			0.03	0.03	0.04	0.04	0.04				
Stably oxidized	(*h*)					0.50	0.58	0.80				
Gold												
Highly polished or film	(*h*)	0.01	0.02	0.03	0.03	0.04	0.05	0.06				
Foil, bright	(*h*)	0.06	0.07	0.07								
Molybdenum												
Polished	(*h*)					0.06	0.08	0.10	0.12	0.15	0.21	0.26
Shot-blasted, rough	(*h*)					0.25	0.28	0.31	0.35	0.42		
Stably oxidized	(*h*)					0.80	0.82					
Nickel												
Polished	(*h*)					0.09	0.11	0.14	0.17			
Stably oxidized	(*h*)					0.40	0.49	0.57				
Platinum												
Polished	(*h*)						0.10	0.13	0.15	0.18		
Silver												
Polished	(*h*)			0.02	0.02	0.03	0.05	0.08				
Stainless steels												
Typical, polished	(*n*)			0.17	0.17	0.19	0.23	0.30				
Typical, cleaned	(*n*)			0.22	0.22	0.24	0.28	0.35				
Typical, lightly oxidized	(*n*)						0.33	0.40				
Typical, highly oxidized	(*n*)						0.67	0.70	0.76			
AISI 347, stably oxidized	(*n*)					0.87	0.88	0.89	0.90			
Tantalum												
Polished	(*h*)								0.11	0.17	0.23	0.28
Tungsten												
Polished	(*h*)							0.10	0.13	0.18	0.25	0.29

Table 13: Emissivity of common materials

7.4 Table: Air properties:

TABLE A–9

Properties of air at 1 atm pressure

Temp. T, °C	Density ρ, kg/m³	Specific Heat c_p J/kg·K	Thermal Conductivity k, W/m·K	Thermal Diffusivity α, m²/s	Dynamic Viscosity μ, kg/m·s	Kinematic Viscosity ν, m²/s	Prandtl Number Pr
−150	2.866	983	0.01171	4.158×10^{-6}	8.636×10^{-6}	3.013×10^{-6}	0.7246
−100	2.038	966	0.01582	8.036×10^{-6}	1.189×10^{-6}	5.837×10^{-6}	0.7263
−50	1.582	999	0.01979	1.252×10^{-5}	1.474×10^{-5}	9.319×10^{-6}	0.7440
−40	1.514	1002	0.02057	1.356×10^{-5}	1.527×10^{-5}	1.008×10^{-5}	0.7436
−30	1.451	1004	0.02134	1.465×10^{-5}	1.579×10^{-5}	1.087×10^{-5}	0.7425
−20	1.394	1005	0.02211	1.578×10^{-5}	1.630×10^{-5}	1.169×10^{-5}	0.7408
−10	1.341	1006	0.02288	1.696×10^{-5}	1.680×10^{-5}	1.252×10^{-5}	0.7387
0	1.292	1006	0.02364	1.818×10^{-5}	1.729×10^{-5}	1.338×10^{-5}	0.7362
5	1.269	1006	0.02401	1.880×10^{-5}	1.754×10^{-5}	1.382×10^{-5}	0.7350
10	1.246	1006	0.02439	1.944×10^{-5}	1.778×10^{-5}	1.426×10^{-5}	0.7336
15	1.225	1007	0.02476	2.009×10^{-5}	1.802×10^{-5}	1.470×10^{-5}	0.7323
20	1.204	1007	0.02514	2.074×10^{-5}	1.825×10^{-5}	1.516×10^{-5}	0.7309
25	1.184	1007	0.02551	2.141×10^{-5}	1.849×10^{-5}	1.562×10^{-5}	0.7296
30	1.164	1007	0.02588	2.208×10^{-5}	1.872×10^{-5}	1.608×10^{-5}	0.7282
35	1.145	1007	0.02625	2.277×10^{-5}	1.895×10^{-5}	1.655×10^{-5}	0.7268
40	1.127	1007	0.02662	2.346×10^{-5}	1.918×10^{-5}	1.702×10^{-5}	0.7255
45	1.109	1007	0.02699	2.416×10^{-5}	1.941×10^{-5}	1.750×10^{-5}	0.7241
50	1.092	1007	0.02735	2.487×10^{-5}	1.963×10^{-5}	1.798×10^{-5}	0.7228
60	1.059	1007	0.02808	2.632×10^{-5}	2.008×10^{-5}	1.896×10^{-5}	0.7202
70	1.028	1007	0.02881	2.780×10^{-5}	2.052×10^{-5}	1.995×10^{-5}	0.7177
80	0.9994	1008	0.02953	2.931×10^{-5}	2.096×10^{-5}	2.097×10^{-5}	0.7154
90	0.9718	1008	0.03024	3.086×10^{-5}	2.139×10^{-5}	2.201×10^{-5}	0.7132
100	0.9458	1009	0.03095	3.243×10^{-5}	2.181×10^{-5}	2.306×10^{-5}	0.7111
120	0.8977	1011	0.03235	3.565×10^{-5}	2.264×10^{-5}	2.522×10^{-5}	0.7073
140	0.8542	1013	0.03374	3.898×10^{-5}	2.345×10^{-5}	2.745×10^{-5}	0.7041
160	0.8148	1016	0.03511	4.241×10^{-5}	2.420×10^{-5}	2.975×10^{-5}	0.7014
180	0.7788	1019	0.03646	4.593×10^{-5}	2.504×10^{-5}	3.212×10^{-5}	0.6992
200	0.7459	1023	0.03779	4.954×10^{-5}	2.577×10^{-5}	3.455×10^{-5}	0.6974
250	0.6746	1033	0.04104	5.890×10^{-5}	2.760×10^{-5}	4.091×10^{-5}	0.6946
300	0.6158	1044	0.04418	6.871×10^{-5}	2.934×10^{-5}	4.765×10^{-5}	0.6935
350	0.5664	1056	0.04721	7.892×10^{-5}	3.101×10^{-5}	5.475×10^{-5}	0.6937
400	0.5243	1069	0.05015	8.951×10^{-5}	3.261×10^{-5}	6.219×10^{-5}	0.6948
450	0.4880	1081	0.05298	1.004×10^{-4}	3.415×10^{-5}	6.997×10^{-5}	0.6965
500	0.4565	1093	0.05572	1.117×10^{-4}	3.563×10^{-5}	7.806×10^{-5}	0.6986
600	0.4042	1115	0.06093	1.352×10^{-4}	3.846×10^{-5}	9.515×10^{-5}	0.7037
700	0.3627	1135	0.06581	1.598×10^{-4}	4.111×10^{-5}	1.133×10^{-4}	0.7092
800	0.3289	1153	0.07037	1.855×10^{-4}	4.362×10^{-5}	1.326×10^{-4}	0.7149
900	0.3008	1169	0.07465	2.122×10^{-4}	4.600×10^{-5}	1.529×10^{-4}	0.7206
1000	0.2772	1184	0.07868	2.398×10^{-4}	4.826×10^{-5}	1.741×10^{-4}	0.7260
1500	0.1990	1234	0.09599	3.908×10^{-4}	5.817×10^{-5}	2.922×10^{-4}	0.7478
2000	0.1553	1264	0.11113	5.664×10^{-4}	6.630×10^{-5}	4.270×10^{-4}	0.7539

Note: For ideal gases, the properties c_p, k, μ, and Pr are independent of pressure. The properties ρ, ν, and α at a pressure P (in atm) other than 1 atm are determined by multiplying the values of ρ at the given temperature by P and by dividing ν and α by P.

Source: Data generated from the EES software developed by S. A. Klein and F. L. Alvarado. Original sources: Keenan, Chao, Keyes, Gas Tables, Wiley, 198; and Thermophysical Properties of Matter, Vol. 3: Thermal Conductivity, Y. S. Touloukian, P. E. Liley, S. C. Saxena, Vol. 11: Viscosity, Y. S. Touloukian, S. C. Saxena, and P. Hestermans, IFI/Plenun, NY, 1970, ISBN 0 30606/020 8.

Table 14: Air properties at 1 atm

7.5 MATLAB Code: For solving the convection coefficient value

```
clear; close; clc

% what we know
T_inf = 22;

T_inf = 22

Bat_H=0.167;

Bat_H = 0.1670

Bat_L=0.1045;

Bat_L = 0.1045

Bat_W=0.0685;

Bat_W = 0.0685

%T_s = 60
k = 0.02528  ;          % W/mK

k = 0.0253

v = 15.34*10^(-6) ;    % kinematics viscosity of the fluid, m^2 /s

v = 1.5340e-05

Pr = 0.7;

Pr = 0.7000


g = 9.81     ;          % gravitational acceleration, m/s^2

g = 9.8100

% characteristic length of the geometry, m
delta_h = Bat_L; %(Bat_H*Bat_W)/(2*(Bat_H+Bat_W))

delta_h = 0.1045

delta_v= Bat_H;

delta_v = 0.1670

%Change in Temperature (number of runs for loop)
count = 150-23;

count = 127

%Set up for arrays to collect data
Temp=zeros(count,1);

Temp =
     0
     0
     0
     0
```


Figure 46: Matlab calculation for solving the convection coefficient

```
      0
      0
      0
      0
      0
      0
      :
      :

CC_H=zeros(count,1);

CC_H =
      0
      0
      0
      0
      0
      0
      0
      0
      0
      0
      :
      :

CC_V=zeros(count,1);

CC_V =
      0
      0
      0
      0
      0
      0
      0
      0
      0
      0
      :
      :

NoBlinds=zeros(count,1);

NoBlinds =
      0
      0
      0
      0
      0
      0
      0
      0
      0
      0
      :
      :

WindowP45=zeros(count,1);

WindowP45 =
      0
      0
      0
```

2

```
         0
         0
         0
         0
         0
         0
         0
         :
         :

WindowN45=zeros(count,1);

WindowN45 =
         0
         0
         0
         0
         0
         0
         0
         0
         0
         0
         :
         :

Window0=zeros(count,1);

Window0 =
         0
         0
         0
         0
         0
         0
         0
         0
         0
         0
         :
         :

ctr=1;

ctr = 1

%Start for loop solve all conditions
for T_s= T_inf+1:1:150;
% the film temp
Tf = ((T_s + T_inf) / 2) + 273.15;

%coefficient of volume expansion, 1/K
B = 1/ Tf;

% Rayleigh number Horizontal Plate
Gr = (g*B*(T_s - T_inf)*delta_h^3) / v^2;
Ra = Gr*Pr;
% Isothermal Horizontal Plate solve Nusselt Number
if (1e+07>Ra)&&(Ra>1e+04) ;
    Nu = 0.54*(Ra)^(1/4);
elseif (1e+11>Ra)&&(Ra>1e+07);
```

3

```matlab
        Nu= 0.15*Ra^(1/3);
end
%Convection coeff. Horizontal Plate
h=(Nu*k)/delta_h;
Temp(ctr)=T_s;
CC_H(ctr)=h; %insert value into array

% Rayleigh number Vertical Plate
Gr = (g*B*(T_s - T_inf)*delta_v^3) / v^2;
Ra = Gr*Pr;
% Isothermal Vertical Plate solve Nusselt Number
if (1e+09>Ra)&&(Ra>1e+04);
Nu_v=0.59*Ra^(0.25);
elseif (1e+13>Ra)&&(Ra>1e+09);
Nu_v=0.1*Ra^(1/3);
end
%Convection coeff. Vertical Plate
h_v=(Nu_v*k)/delta_v;
CC_V(ctr)=h_v; %insert value into array


DeltaTemp=(T_s)-22;
%Solve Study equations based on room condictions
NoBlind_h=1.74*(DeltaTemp^(0.33));
NoBlinds(ctr)=NoBlind_h;
WindowP45(ctr)=1.89*(DeltaTemp^(0.25));
WindowN45(ctr)=1.67*(DeltaTemp^(0.25));
Window0(ctr)=1.48*(DeltaTemp^(0.25));

ctr=ctr+1; %For Loop counter
end

C = [Temp CC_H CC_V NoBlinds WindowP45 WindowN45 Window0]

C =
    23.0000    2.3932    2.3256    1.7400    1.8900    1.6700    1.4800
    24.0000    2.8448    2.7644    2.1872    2.2476    1.9860    1.7600
    25.0000    3.1470    3.0581    2.5003    2.4874    2.1978    1.9478
    26.0000    3.3802    3.2847    2.7493    2.6729    2.3617    2.0930
    27.0000    3.5726    3.4717    2.9594    2.8262    2.4972    2.2131
    28.0000    3.7377    3.6321    3.1430    2.9580    2.6137    2.3163
    29.0000    3.8829    3.7732    3.3070    3.0742    2.7164    2.4073
    30.0000    4.0130    3.8997    3.4560    3.1786    2.8086    2.4891
    31.0000    4.1312    4.0145    3.5929    3.2736    2.8925    2.5634
    32.0000    4.2397    4.1200    3.7201    3.3609    2.9697    2.6319
        :
        :


disp(C)

    23.0000    2.3932    2.3256    1.7400    1.8900    1.6700    1.4800
    24.0000    2.8448    2.7644    2.1872    2.2476    1.9860    1.7600
    25.0000    3.1470    3.0581    2.5003    2.4874    2.1978    1.9478
    26.0000    3.3802    3.2847    2.7493    2.6729    2.3617    2.0930
    27.0000    3.5726    3.4717    2.9594    2.8262    2.4972    2.2131
    28.0000    3.7377    3.6321    3.1430    2.9580    2.6137    2.3163
    29.0000    3.8829    3.7732    3.3070    3.0742    2.7164    2.4073
    30.0000    4.0130    3.8997    3.4560    3.1786    2.8086    2.4891
```

4

31.0000	4.1312	4.0145	3.5929	3.2736	2.8825	2.5634
32.0000	4.2397	4.1200	3.7201	3.3609	2.9697	2.6319
33.0000	4.3401	4.2176	3.8389	3.4420	3.0413	2.6953
34.0000	4.4337	4.3085	3.9507	3.5177	3.1082	2.7546
35.0000	4.5215	4.3938	4.0565	3.5888	3.1710	2.8103
36.0000	4.6041	4.4741	4.1560	3.6559	3.2303	2.8628
37.0000	4.6823	4.5501	4.2526	3.7195	3.2865	2.9126
38.0000	4.7565	4.6222	4.3442	3.7800	3.3400	2.9600
39.0000	4.8271	4.6908	4.4320	3.8377	3.3910	3.0052
40.0000	4.8946	4.7564	4.5164	3.8930	3.4398	3.0485
41.0000	4.9592	4.8191	4.5977	3.9459	3.4866	3.0899
42.0000	5.0211	4.8793	4.6762	3.9969	3.5316	3.1298
43.0000	5.0807	4.9372	4.7521	4.0459	3.5750	3.1682
44.0000	5.1380	4.9929	4.8256	4.0932	3.6168	3.2053
45.0000	5.1933	5.0466	4.8969	4.1390	3.6572	3.2411
46.0000	5.2467	5.0985	4.9661	4.1833	3.6963	3.2758
47.0000	5.2984	5.1487	5.0335	4.2262	3.7342	3.3094
48.0000	5.3484	5.1973	5.0991	4.2678	3.7710	3.3420
49.0000	5.3969	5.2445	5.1630	4.3083	3.8068	3.3737
50.0000	5.4440	5.2903	5.2253	4.3476	3.8415	3.4045
51.0000	5.4897	5.3347	5.2862	4.3859	3.8754	3.4345
52.0000	5.5342	5.3779	5.3456	4.4233	3.9084	3.4637
53.0000	5.5775	5.4200	5.4038	4.4597	3.9406	3.4922
54.0000	5.6197	5.4610	5.4607	4.4952	3.9720	3.5201
55.0000	5.6609	5.5010	5.5164	4.5299	4.0026	3.5472
56.0000	5.7010	5.5400	5.5711	4.5638	4.0326	3.5738
57.0000	5.7402	5.5780	5.6246	4.5970	4.0619	3.5998
58.0000	5.7784	5.6152	5.6771	4.6295	4.0906	3.6252
59.0000	5.8158	5.6516	5.7287	4.6614	4.1188	3.6502
60.0000	5.8524	5.6871	5.7793	4.6925	4.1463	3.6746
61.0000	5.8882	5.7219	5.8291	4.7231	4.1733	3.6985
62.0000	5.9232	5.7559	5.8780	4.7531	4.1998	3.7220
63.0000	5.9575	5.7893	5.9261	4.7825	4.2258	3.7451
64.0000	5.9911	5.8219	5.9734	4.8114	4.2514	3.7677
65.0000	6.0241	5.8540	6.0200	4.8398	4.2765	3.7899
66.0000	6.0564	5.8854	6.0658	4.8677	4.3011	3.8118
67.0000	6.0882	5.9162	6.1110	4.8951	4.3253	3.8332
68.0000	6.1193	5.9465	6.1554	4.9221	4.3492	3.8544
69.0000	6.1499	5.9762	6.1993	4.9486	4.3726	3.8751
70.0000	6.1799	6.0054	6.2425	4.9748	4.3957	3.8956
71.0000	6.2094	6.0341	6.2852	5.0005	4.4184	3.9157
72.0000	6.2384	6.0622	6.3272	5.0258	4.4408	3.9355
73.0000	6.2669	6.0900	6.3687	5.0507	4.4628	3.9551
74.0000	6.2950	6.1172	6.4096	5.0753	4.4845	3.9743
75.0000	6.3226	6.1440	6.4500	5.0995	4.5059	3.9933
76.0000	6.3497	6.1704	6.4899	5.1234	4.5270	4.0120
77.0000	6.3764	6.1964	6.5293	5.1470	4.5479	4.0304
78.0000	6.4028	6.2219	6.5683	5.1702	4.5684	4.0486
79.0000	6.4287	6.2471	6.6068	5.1931	4.5887	4.0666
80.0000	6.4542	6.2719	6.6448	5.2158	4.6086	4.0843
81.0000	6.4793	6.2963	6.6824	5.2381	4.6284	4.1018
82.0000	6.5041	6.3204	6.7195	5.2602	4.6479	4.1191
83.0000	6.5285	6.3442	6.7563	5.2819	4.6671	4.1361
84.0000	6.5526	6.3676	6.7926	5.3035	4.6861	4.1530
85.0000	6.5764	6.3906	6.8286	5.3247	4.7049	4.1696
86.0000	6.5998	6.4134	6.8642	5.3457	4.7235	4.1861
87.0000	6.6229	6.4358	6.8994	5.3665	4.7418	4.2023
88.0000	6.6457	6.4580	6.9342	5.3870	4.7600	4.2184
89.0000	6.6682	6.4798	6.9687	5.4073	4.7779	4.2343
90.0000	6.6904	6.5014	7.0029	5.4274	4.7956	4.2500
91.0000	6.7123	6.5227	7.0367	5.4472	4.8131	4.2655
92.0000	6.7339	6.5437	7.0702	5.4668	4.8305	4.2809
93.0000	6.7551	6.5645	7.1034	5.4863	4.8476	4.2961
94.0000	6.7764	6.5850	7.1362	5.5055	4.8646	4.3112
95.0000	6.7972	6.6053	7.1688	5.5245	4.8814	4.3261

5

```
 96.0000   6.8178   6.6253   7.2010   5.5433   4.8981   4.3408
 97.0000   6.8382   6.6450   7.2330   5.5620   4.9145   4.3554
 98.0000   6.8583   6.6646   7.2647   5.5804   4.9308   4.3698
 99.0000   6.8781   6.6839   7.2961   5.5987   4.9470   4.3841
100.0000   6.8978   6.7030   7.3272   5.6168   4.9630   4.3983
101.0000   6.9172   6.7218   7.3581   5.6347   4.9788   4.4123
102.0000   6.9364   6.7405   7.3887   5.6524   4.9945   4.4262
103.0000   6.9554   6.7589   7.4191   5.6700   5.0100   4.4400
104.0000   6.9741   6.7772   7.4492   5.6874   5.0254   4.4536
105.0000   6.9927   6.7952   7.4790   5.7047   5.0406   4.4672
106.0000   7.0111   6.8131   7.5086   5.7218   5.0558   4.4806
107.0000   7.0292   6.8307   7.5380   5.7387   5.0707   4.4938
108.0000   7.0472   6.8482   7.5672   5.7555   5.0856   4.5070
109.0000   7.0650   6.8655   7.5961   5.7722   5.1003   4.5200
110.0000   7.0826   6.8826   7.6248   5.7887   5.1149   4.5330
111.0000   7.1000   6.8995   7.6533   5.8051   5.1284   4.5458
112.0000   7.1173   6.9163   7.6816   5.8213   5.1437   4.5585
113.0000   7.1343   6.9329   7.7096   5.8374   5.1579   4.5711
114.0000   7.1512   6.9493   7.7375   5.8534   5.1721   4.5836
115.0000   7.1680   6.9655   7.7651   5.8692   5.1861   4.5960
116.0000   7.1845   6.9816   7.7926   5.8850   5.1999   4.6083
117.0000   7.2009   6.9976   7.8198   5.9006   5.2137   4.6205
118.0000   7.2172   7.0134   7.8469   5.9160   5.2274   4.6327
119.0000   7.2333   7.0290   7.8738   5.9314   5.2409   4.6447
120.0000   7.2492   7.0445   7.9005   5.9466   5.2544   4.6566
121.0000   7.2650   7.0598   7.9270   5.9617   5.2678   4.6684
122.0000   7.2806   7.0750   7.9533   5.9767   5.2810   4.6802
123.0000   7.2961   7.0901   7.9795   5.9916   5.2942   4.6918
124.0000   7.3115   7.1050   8.0055   6.0064   5.3072   4.7034
125.0000   7.3267   7.1198   8.0313   6.0210   5.3202   4.7149
126.0000   7.3418   7.1344   8.0569   6.0356   5.3330   4.7263
127.0000   7.8329   7.1489   8.0824   6.0501   5.3458   4.7376
128.0000   7.8533   7.1633   8.1077   6.0644   5.3585   4.7488
129.0000   7.8748   7.1776   8.1329   6.0787   5.3711   4.7600
130.0000   7.8955   7.1917   8.1579   6.0928   5.3836   4.7711
131.0000   7.9168   7.2057   8.1828   6.1069   5.3960   4.7821
132.0000   7.9363   7.2196   8.2075   6.1208   5.4083   4.7930
133.0000   7.9565   7.2334   8.2320   6.1347   5.4206   4.8039
134.0000   7.9766   7.2470   8.2564   6.1485   5.4328   4.8147
135.0000   7.9964   7.2606   8.2807   6.1621   5.4449   4.8254
136.0000   8.0162   7.2740   8.3048   6.1757   5.4569   4.8360
137.0000   8.0357   7.2873   8.3287   6.1892   5.4688   4.8466
138.0000   8.0551   7.3005   8.3526   6.2025   5.4806   4.8571
139.0000   8.0744   7.3136   8.3763   6.2160   5.4924   4.8675
140.0000   8.0935   7.3266   8.3998   6.2292   5.5041   4.8779
141.0000   8.1125   7.3395   8.4233   6.2424   5.5157   4.8882
142.0000   8.1314   7.3523   8.4465   6.2554   5.5273   4.8984
143.0000   8.1501   7.3649   8.4697   6.2684   5.5388   4.9086
144.0000   8.1686   7.3775   8.4927   6.2813   5.5502   4.9187
145.0000   8.1871   7.3900   8.5157   6.2942   5.5616   4.9288
146.0000   8.2054   7.4024   8.5384   6.3069   5.5728   4.9388
147.0000   8.2235   7.4147   8.5611   6.3196   5.5840   4.9487
148.0000   8.2416   7.4269   8.5836   6.3322   5.5951   4.9585
149.0000   8.2595   7.4390   8.6061   6.3447   5.6062   4.9684
150.0000   8.2772   7.4510   8.6284   6.3572   5.6172   4.9781
```

```matlab
figure
plot(Temp,CC_H,Temp,CC_V,Temp,NoBlinds,Temp,WindowP45,Temp,WindowN45,Temp,Window0);
title('Change of Convection Coef. Vs Temp.');
legend({'Horizonal Plate','Vertical Plate','No Blinds','Window +45 Blinds','Window -45 Blinds',
xlabel('Temperature C') ;
ylabel('Convection Coefficient') ;
```

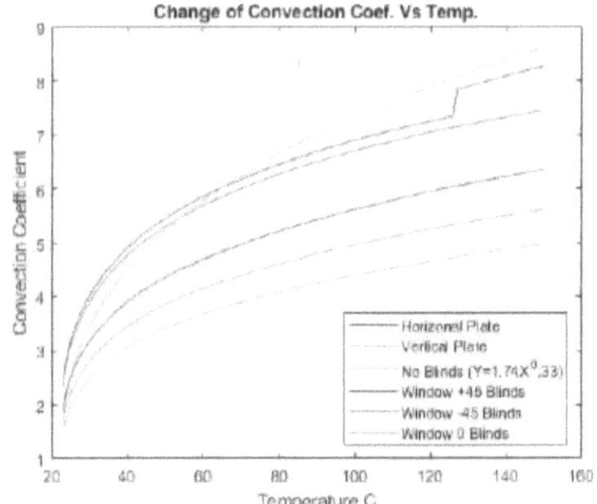


```
filename= 'ConvectionCoe.xlsx';

filename =
'ConvectionCoe.xlsx

xlswrite(filename,C);
```

7

93

7.6 MATLAB code: Specific heat

```matlab
Al_m = 0.310197357; % Mass of Aluminum
SH_Al = 900; % Specific Heat


Plastic_m = 0.182626756; % Mass of Plastic Holder kg
SH_Pl = 1500; % Specific Heat


Silicone_m = 0.17765646; % Mass Silicone Sleeve kg
SH_Sil = 1075; % Specific Heat


Battery_m = 1.864; % Mass of all battery cells kg
SH_Bat = 960; % Specific Heat


t = 466-1; % Time in seconds
ctr = 1;
Total_Heat_Power=zeros(t,1);
Time=zeros(t,1);
Ti = 22.4 ;        % Initial Temp
for t=1:1:466

    Al_temp = 0.1308*t+21.169; % Temperature increase over time 35 Amp
        Al_HP = ((Al_temp-Ti)*Al_m*SH_Al)/t; % Heat energy required to raise temp. of Aluminum


    Plastic_temp = 0.1215*t+19.953; % Temperature increase over time 35 Amp
    Plastic_HP = ((Plastic_temp-Ti)*Plastic_m*SH_Pl)/t; % Heat energy required to raise temp. 


    Silicone_temp = 0.1219*t+21.506; % Temperature increase over time 35 Amp
    Silicone_HP = ((Silicone_temp-Ti)*Silicone_m*SH_Sil)/t; % Heat energy required to raise ten

    Battery_Temp = 0.1268*t+23.193; % Temperature increase over time 35 Amp
    Battery_HP = ((Battery_Temp-Ti)*Battery_m*SH_Bat)/t; % Heat energy required to raise temp.

    Total_Heat_Power(ctr)=(Al_HP+Plastic_HP+Silicone_HP+Battery_HP)/40; %Total heat power per 
    Time(ctr)=t;

    ctr=ctr+1 ;
end

C = [Time Total_Heat_Power] ;
disp(C)

    1.0000    13.8568
    2.0000    10.9282
    3.0000     9.9520
    4.0000     9.4639
    5.0000     9.1710
    6.0000     8.9758
    7.0000     8.8363
    8.0000     8.7317
```

1

Figure 47:Matlab code for calculating the heat power required to heat up the material inside the battery pack.

```
464.0000    8.0122
465.0000    8.0121
466.0000    8.0121


filename= 'Heat_Power_SH.xlsx'

filename =
'Heat_Power_SH.xlsx'

xlswrite(filename,C);
```

7.7 MATLAB code: Calculating State Of Charge (SOC)

```
clear,clc

SOCi = 100 % State of charge

SOCi = 100


i= 35 %Discharge Current

i = 35


t= 482 % Time in seconds for discharge

t = 482


Cap= 6 % Amp Hours of battery pack

Cap = 6

Capsec = Cap*60*60 % Amp Seconds of the battery pack

Capsec = 21600


SOCf= ((Capsec-(i*t))/Capsec)*100 % Final State of Charge

SOCf = 21.8981
```

Figure 48:Matlab code for calculating the SOC.

9 783384 241382